Using Finite Elements in Mechanical Design

Using Finite Elements in Mechanical Design

Dr. J. Toby Mottram and
Dr. Christopher T. Shaw

McGRAW-HILL BOOK COMPANY

London · New York · St Louis · San Francisco · Auckland · Bogotá · Caracas
Lisbon · Madrid · Mexico · Milan · Montreal · New Delhi · Panama · Paris
San Juan · São Paulo · Singapore · Sydney · Tokyo · Toronto

Published by
McGRAW-HILL Book Company Europe
Shoppenhangers Road, Maidenhead, Berkshire, SL6 2QL, England
Telephone 01628 23432
Facsimile 01628 770224

British Library Cataloguing in Publication Data
Mottram, J. Toby
  Using finite elements in mechanical design
  1. Finite element method 2. Mechanical engineering – Data processing
  3. Design, Industrial 4. Structural design – Data processing 5. Computer-aided
  design 6. Mechanics, Analytic
  I. Title II. Shaw, Christopher T.
  620'.0042'0285

  ISBN 0-07-709093-4

Library of Congress Cataloging-in-publication Data
  Mottram, J. Toby (James Toby),
    Using finite elements in mechanical design / J. Toby Mottram and
  Christopher T. Shaw.
      p. cm.
    Includes index.
    ISBN 0-07-709093-4
    1. Finite element method–Data processing. 2. Engineering design–
  Data processing. 3. Machine design–Data processing. 4. Computer
  -aided design.
  I. Shaw, Christopher T. (Christopher Thomas). II. Title.
  TA347.F5M68 1996
  620.1'05'0285–dc20
                                                     95-46042 CIP

McGraw-Hill

A Division of The **McGraw·Hill** Companies

1234 CUP 99876

Typeset by Keyword Typesetting Services Ltd, Wallington, Surrey
Printed and bound in Great Britain at the University Press, Cambridge

Printed on permanent paper in compliance with ISO Standard 9706

This book is dedicated:

by JTM to his wife, Liz, and sons, Thomas and Jamie

by CTS to the memory of his father, Albert

CONTENTS

MOTIVATION

This book is aimed at those who are new to using finite element software for mechanical or structural design. There are many textbooks that explain the theory behind the use of finite elements in this context, but few that delve into the subtlety of the modelling process itself. While an understanding of the theory is vital to the successful use of finite elements, emphasis will be given here to the practicalities of modelling. As well as describing the modelling process, examples—both simple test cases and real engineering problems—will be solved.

The text will consider in detail only the linear elastic small deformation behaviour of structures. This restriction is necessary to limit the length of the book, but it is justified as much of the analysis carried out by industry falls squarely within this category. Some insight is given into more advanced analysis techniques at the end of the book.

OBJECTIVES

After reading the material, a reader will be able to:

- understand the ways in which mechanical structures and devices transmit the loads applied to them
- appreciate the historical development of the finite element method
- appreciate some of the theoretical background to the fundamental equations that govern stress, strain and displacements
- follow the analysis process for a design
- model the geometry of a design on a computer
- define the mechanical properties of materials used in design (an open-ended problem similar to the modelling of turbulence in fluids)
- begin to use finite element software to predict stresses and displacements within designs for given applied loads
- have an understanding of the computer technology required to carry out an analysis.

ORGANIZATION

Chapter 1 looks at the design process, how designs are evaluated, the use of computers in this process and a historical review of the development of the finite element method. Chapter 2 then develops the background to continuum mechanics as it relates to the transmission of forces through a solid material, developing the governing equations for equilibrium, strain and compatibility as well as looking at material properties. Chapter 3 discusses the use of finite elements in producing numerical analogues to the equations that can be solved on a computer. After this more technical material, the main core of the text, follows as subsequent chapters look at the use and acquisition of the necessary hardware and software and at the process of using the finite element method to obtain design information. This is done by looking at this process in a series of stages: thinking about the design problem; creating a computer model of the geometry of the design; building a finite element mesh; applying boundary conditions; obtaining a solution; and analysing the results. Finally, Chapter 10 looks at advanced topics.

Throughout the text SI units have been used, with the exception that mm have been used for length in places. This useage should be clear when the text is read.

USING THE BOOK

This book is intended to be both an introductory guide and a working reference. Students at both undergraduate and postgraduate level who are studying courses on finite element methods may want to read the chapters in sequence, but those who are modelling may just at some stage want information on a particular aspect of the process of evaluating designs. For these readers we have tried to keep the individual chapters as self-contained as possible so that they may where necessary be read in isolation. We have also attempted to keep the mathematics to a minimum and to keep the introduction to the finite element method as general as possible so that a wide range of people from student to practising engineer will find the book of use.

We hope that readers will find this book a suitable bridge between the complementary areas of continuum mechanics, numerical analysis and computing that together enable computers to be used to evaluate designs. If some of the pain that is currently found when building this bridge is reduced then our effort will have been worth while.

Note that to keep production costs to a minimum, only monochrome figures have been used.

FURTHER STUDY

For some readers the presentation here will suffice, but others may want to learn more about processes both in mathematical and physical terms. For those who want to continue their learning into structural behaviour, several ways are open. The first is simply to observe structures and think about why they deform as they do, or why they are designed as they are. The second is to read other textbooks, which can be of great help. Appropriate references are given throughout this book to suitable texts. Finally, courses are often run by universities in the field of structural mechanics, and it may be that one of these may provide the necessary insight.

ACKNOWLEDGEMENTS

Both authors are grateful to the professional help given by staff at McGraw-Hill during the development of the material and to the reviewers who provided such helpful comments. Encouragement has come from many places and this book has been enriched by the examples provided by both contacts in industry and students. In particular, we are grateful to Michael Bradley for developing the model of the bicycle frame as a third-year undergraduate project, to Geriant Williams and Ralph Smith at Jaguar for providing the con-rod example, to William Turner and Andy Ellis at the Motor Industry Research Association for the vehicle body shell example and to Martyn Pinfold of the Warwick Manufacturing Group, University of Warwick and Geoff Calvert of the Rover Group, for the composite suspension arm.

Thanks are also owed to Prentice Hall for permission to use a modified form of material that has already appeared in Shaw (1992). Indeed, they actively encouraged this use!

Finally, thanks are owed to SDRC, and in particular our contact there, Andy Burton, for the provision of the I-DEAS software at the educational rate and for the encouragement given when this book was just a figment of our imagination.

Dr. J. Toby Mottram and Dr. Christopher T. Shaw
University of Warwick

ITALIC SYMBOLS

a	side length of quadrilateral element, mm
a_i	generalized coefficients in Rayleigh–Ritz method
A	area, mm^2
b	side length of quadrilateral element, mm
C_{ij}	elastic stiffnesses, N mm^{-2}
d	diameter of disk, lever arm, mm
D	diameter of bar, mm; objective function
E	Young's modulus, kN mm^{-2}
F	resultant force, N
G	shear modulus, kN mm^{-2}
h	length of real element, mm
I_a	second moment of area, mm^4
I_p	polar second moment of area, mm^4
J	determinant of [$\mathbf{J}$] (the Jacobian)
J_Γ	boundary Jacobian
k	elastic stiffness, N mm^{-1}
K_s	shear correction factor
K_{ij}	stiffness term, i is node number, j is nodal displacement
l, m, n	directional cosines
L	length, beam element, beam span, mm
m	mass, kg
M	moment, N mm; total number of elements
n	number of degrees of freedom; number of Gauss points in numerical integration schemes
N_i	shape function terms in matrix [$\mathbf{N}$]
O	order, e.g. $O(h^2)$ is a term of order h^2
P	line load, N; number of concentrated load components
q	pressure load per unit length, N mm^{-1} per unit length
Q	heat source, °C mm^{-2}
r	radius, mm
R	radius of curvature, mm
S	surface area, mm^2
S_{ij}	elastic compliances, N^{-1} mm^2
t	thickness of plate, mm; time, s
T	temperature, °C; torque, N mm

u, v, w	displacement components in x-, y-, z-directions, respectively, mm
V	volume, mm^3
w_{nl}, w_{nk}	weighted factors in numerical integration scheme
x, y, z	Cartesian coordinate system
X, Y, Z	body force components per unit volume in x-, y-, z-directions respectively, $N\,mm^{-3}$
$\bar{X}, \bar{Y}, \bar{Z}$	surface traction components in x-, y-, z-directions respectively, $N\,mm^{-2}$

GREEK SYMBOLS

α_i	generalized coefficient(s)
γ	shear strain (e.g. $\gamma_{xy}, \gamma_{yz}, \gamma_{xz}$)
δ	small bit of, finite element degree of freedom
Δ	change in length, mm, displacements due to external concentrated loads
Δt	increment in time, s
ε	direct strain (e.g. $\varepsilon_x, \varepsilon_y, \varepsilon_z$)
η	isoparametric coordinate, damping factor(s)
θ	angle of twist per unit length, rotational degrees of freedom, rad; weighting factor
ν	Poisson's ratio
ξ	isoparametric coordinate, damping factor
ξ_{nk}, η_{nl}	sample points in numerical integration scheme
Π_p	total potential energy functional
ρ	mass density, $kg\,m^{-3}$
σ	direct stress (e.g. $\sigma_x, \sigma_y, \sigma_z$), $N\,mm^{-2}$
τ	shear stress (e.g. $\tau_{xy}, \tau_{yz}, \tau_{xz}$), $N\,mm^{-2}$
Φ	Airy's stress function
ω	natural frequency, Hz; relaxation factor

SUBSCRIPTS AND SUPERSCRIPTS

a	axial loading
b	bending
d	damped
e	element
e	element number
i, j, k, l	labels for nodes
m, n, o, p	labels for nodes
max	maximum
min	minimum
n	number of degrees of freedom, number of simultaneous equations

nat	natural
new	new value
next	next value
old	old value
O	initial value
p	concentrated load component
s	shear
U	ultimate, upper
vonMises	von Mises' stress
w	web
x, y, z	Cartesian coordinates
Y	yield
1,2,3	principal values

MATHEMATICAL SYMBOLS

[ ]	rectangular or square matrix
{ }	vector or column matrix
[ ]^T	matrix transpose
[ ]$^{-1}$	matrix inverse
∂	partial differential
d	total differential
$\{\mathbf{a}\}$	coefficients in Rayleigh–Ritz method
$\{\mathbf{b}\}$	general vector
$[\mathbf{A}]$	Vandermode matrix, general square matrix, system matrix
$[\mathbf{B}]$	spatial derivative(s) of the field variable(s) are$[\mathbf{B}]\{\boldsymbol{\delta}^e\}$
$[\mathbf{C}]$	damping matrix
$[\mathbf{D}]$	material stiffness(es) matrix
$[\mathbf{f}]$	displacement function(s) matrix
$\{\mathbf{F}^e\}$	element force vector $=\{\mathbf{F_q}^e\} + \{\mathbf{R}^e\}$
$\{\mathbf{F}\}$	global force vector
$\{\mathbf{F_q}^e\}$	consistent element force vector due to distributed loading
$\{\mathbf{F}(t)\}$	forcing function vector
$[\mathbf{I}]$	unit matrix
$[\mathbf{J}]$	the Jacobian matrix
$[\mathbf{k}]$	element stiffness matrix
$[\mathbf{K}]$	global or structure stiffness matrix
$[\mathbf{L}]$	lower triangular matrix
$[\mathbf{M}^e]$	element mass matrix
$[\mathbf{M}]$	mass matrix
$[\mathbf{N}]$	element shape function matrix
$\{\mathbf{r}\}$	residual errors vector
$\{\mathbf{R}\}$	vector of applied load(s) corresponding to displacement components

$\{\mathbf{u}\}$	displacement components
$[\mathbf{U}]$	upper triangular matrix
$\{\mathbf{x}\}$	general vector
$\{\mathbf{X}\}$	body force(s) per unit volume vector
$\{\bar{\mathbf{X}}\}$	surface traction(s) vector
$\{\alpha\}$	generalized coefficient vector
$\{\boldsymbol{\delta}\}$	global nodal degree of freedom vector
$\{\boldsymbol{\delta}^{\mathbf{e}}\}$	element nodal degree of freedom vector
$\{\bar{\boldsymbol{\delta}}\}$	nodal amplitude(s) vector
$\left\{\dfrac{\partial \boldsymbol{\delta}}{\partial t}\right\}$	velocity vector of degrees of freedom
$\left\{\dfrac{\partial^2 \boldsymbol{\delta}}{\partial t^2}\right\}$	acceleration of degrees of freedom
$\{\Delta_p\}$	displacement(s) associated with P external concentrated load(s) vector
$\{\boldsymbol{\varepsilon}\}$	strain vector
$\{\boldsymbol{\sigma}\}$	stress vector
$[\boldsymbol{\partial}]$	partial differential operator matrix

1

RELATIONSHIP BETWEEN DESIGN
AND FINITE ELEMENTS

In mechanical design some form of strength analysis is usually required as part of the design process. Traditionally this has been done by simple engineering calculations, but as product performance becomes more important and as designs become more complex so these simple methods have become inadequate. With the increase in computational capacity and the availability of software that can predict loading for complex geometries and material behaviour so there has been an explosion in the use of the finite element method both in academia and in industry. To understand this growth and its implication for design, it is necessary to look at what the design process is all about, with particular reference to the field of mechanical engineering. By looking at the design process, it is possible to look at how design solutions are evaluated and at where computer technology fits into this process. Then the historical development of the finite element method can be discussed, to illustrate the benefits of using this technology.

From looking at the design process it will become clear that to appreciate fully both the background theory of the finite element method and the ways in which it is used, analysts must have some understanding of the mechanics that structures use to transmit loads. These mechanics can then be translated to a mathematical form suitable for solution by computer. Chapter 2 develops the understanding and Chapter 3 explains the translation before the remaining chapters show how the finite element analysis is carried out in practice.

1.1 THE DESIGN PROCESS

All around there are examples of artefacts that have been designed with their mechanical properties in mind. For example, computers have a structure that must be capable of transmitting the weight of the cathode ray tube and its associated electromagnets through to the desk. While doing this the structure must also not deform too much. Equally, the disk drive attached to the computer must be able to withstand the loads imposed by the spinning disk and the moving heads that read the data. Regardless of the product being considered, the process undertaken by its designers from first thoughts to final design will be similar.

There has been much research into the stages that are followed during the design process, but many people describe the process in similar terms. In particular, the key stages in the process (see Fig. 1.1) can be identified as follows:

- *Recognition of a need*, which may well come from market research or a request to tender in terms of a design brief for a new or modified product or from other sources such as the simple intuition of a designer.
- *Definition of the problem or specification,* where, from the starting point of a design brief or whatever, a full technical specification of the desired product must be developed through consultation between designers and those who want the product, for example those who sell into the market-place or the end-

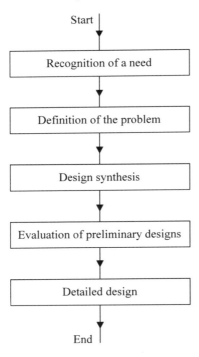

Figure 1.1 The design process.

users. Such a specification should contain all the relevant technical information describing the product, covering a variety of design considerations such as function, materials, appearance, environmental effect, product life, reliability, safety, interchangeability, standardization of parts, maintenance and service requirements and costs, together with any constraints that the design must meet. Further, the criteria that the design is to meet must be detailed. It is worth noting here that as well as essential items any desirable items must be specified too.

- *Design synthesis*, during which a variety of solutions are generated as rough ideas. Here the geometry of the product is most important and can be determined not only by the function of the product but also by the manufacturing process or the material to be used.
- *Evaluation of preliminary designs*, when each design is tested in some sense against the specification. For example, this analysis of a design might include simple physical testing or computer modelling, and should enable unsuitable designs to be filtered out. Also optimization might be achieved through a review of the designs, such that at the end of this stage a design that has the capacity to meet the specification in full and is best in some sense should be available.
- *Detailed design*, when the winning design is specified in full, i.e. in terms of size, materials, tolerances and shape, ready for manufacture.

A simpler version of this might consist of just four phases; clarification, concept design, embodiment design and detail design, but, regardless of the words used to describe the process, three key activities can be seen to be carried out by designers. These are:

- *generation* of solutions to the design problem
- *evaluation* of solutions to see if they meet the design specification
- *communication* of solutions, traditionally in the form of drawings, to many people such as customers and manufacturers.

In Fig. 1.1 the process progresses neatly from stage to stage. In reality, however, it is an inherent feature of the process that some form of iteration will take place. For example, problems might be found during the later stages because there is a lack of definition within the specification. If this lack of definition did not exist then there would be little or no scope for engineers and designers to be ingenious in generating solutions. However, it might mean that a review of the design specification is necessary and so the process loops back towards the start again. Also, none of the design solutions may meet the specification and so a rethink may be necessary.

When several designs are found that meet the specification the process changes into the search for the so-called optimal design. This is the one that is seen as best in some sense. Again iteration might be required to refine the best design still further.

It is worth noting that this might be considered a traditional view of the design process, as recent thinking considers the embodiment of the design process within the whole product life cycle. This is known as *simultaneous* or *concurrent* engineering, where the design process takes into account not only the market into which the product must be launched, but also the manufacture of the subassemblies of the product and the final assembly. By doing this costly mistakes should be avoided and the product lead time reduced.

1.2 EVALUATING DESIGNS

In the evaluation process, all designs have to be checked for conformance with the specification. Clearly the specification has to be sufficiently detailed for the evaluation process to determine which designs meet the specification and which do not.

The list of areas covered in a typical specification indicates that various evaluation techniques may need to be used. In particular, the function category may include:

- strength under load
- displacement of the structure under load
- thermal behaviour
- fluid flow properties
- electrical/electronic behaviour.

All of these can be described as properties of the physical behaviour of the design. This book, however, is about the use of finite element methods in mechanical design—the design of the physical structure to carry an applied load or loads. Consequently the text is concerned with the use of continuum mechanics in determining such things as

- the strength of a structure
- the displacement of parts of the structure when loaded
- the effect of heat on internal stresses and displacement
- the optimal (minimum) thicknesses of material for a given displacement
- the fatigue life of a structure (or product life)
- the dynamic response of a structure
- crash worthiness.

1.3 USING COMPUTERS FOR DESIGN EVALUATION

For computer technology to be adopted at any stage of the design process there must be some benefit to the process as a whole. As the acquisition of the technology is fairly expensive these benefits have to be compared to the investment. For example, the following skills are required by a designer or engineer:

- *Analysis* Where physical or numerical experiments are carried out and the results are used for decision-making.

- *Reasoning* Where logic and intellect are used to produce a reasoned argument for following certain decisions. This is often developed through experience in engineering design.
- *Storage and handling of data* Where information is processed, both as input to a system and as output.
- *Error handling* Where errors to a given tolerance must be detected and corrected.

By adding computer technology to the armoury of the designer, the best human qualities of the designer can be linked by the best qualities of the computer. For example, humans excel at reasoning, developed from experience, error handling and the practical aspects of analysis. Equally, computers are far better at the storage and handling of data and the numerical aspects of analysis where repetitive calculations have to be made. The combination of these two, i.e. designers who use modern computer technology, should be able to produce design solutions more quickly and more accurately than would be possible if designers were to work without computers.

In summary, the benefits of using computer-aided design tools are as follows:

- *increased design efficiency and effectiveness* because repetitive tasks are carried out by the computer
- *simplification of the design process* by using an integrated data storage system, allowing many people access to relevant information when they want it
- *economy of material and labour* through a reduction in the amount of prototype building and testing required
- *better documentation* through computer-generated drawings, bills of materials, parts lists, work schedules and so on

1.4 TYPICAL DESIGN SITUATIONS

Section 1.2 listed some typical properties of a structure that might be of interest to a designer. In both aircraft and road vehicle design these properties need to be known before the first prototype flies or is driven on the road. Looking at each in turn:

- *The strength of a structure* needs to be known so that failure loads can be predicted. It is no use having an aircraft that cannot support its own weight when fully laden with fuel and passengers. Consequently calculations are carried out to predict the internal stresses in the wing structure when the aircraft is generating sufficient lift for certain manoeuvres plus an additional safety factor of load. At this condition the internal stresses should not lead to any failure of the material.
- *The displacement of parts of the structure when loaded* needs to be known. For example, the deflection of the wing tips of an aircraft when the wing is full of fuel will determine, in part, the design of the undercarriage.

- *The effect of heat on internal stresses and displacement* is particularly important in the area around the engine or engines. Hence, the displacements and loads due to the heating of the structure both of the engine itself and the surrounding vehicle body need to be determined.
- *The optimal (minimum) thicknesses of material for a given displacement* can be crucial to the handling characteristics of a road vehicle. For example, when designing an open-top version of a saloon car the removal of the roof makes the body shell much more flexible. Engineers need to know how much extra material will be needed in others parts of the structure to return the body shell to an acceptable level of stiffness. By calculating the displacements of some extremity of the shell under a given load for various configurations, the stiffness of the shell can be predicted as a function of the material added.
- *The fatigue life of a structure (or product life)* can be predicted by calculating the mean stress in the structure and the operating stress range seen at a known frequency. This might occur, for example, when an aircraft lands and standard design tables give the life of the structure before cracking initiates.
- *The dynamic response of a structure* is calculated to enable any resonances of the structure to be determined. This occurs if the structure is excited at frequencies close to the natural frequencies of the structure.
- *Crash worthiness* is particularly important for road vehicles and, in fact, a standard crash test must be carried out on at least one vehicle of each type before approval is given for the vehicles to be sold to the public. Predictions of the behaviour of the body shell are routinely made to assist in the design of the vehicle substructure that must absorb the energy of impact.

From these examples it can be seen that the ways in which mechanical designs carry and transmit applied loads to the supports are very important and need to be considered at an early stage in the design process. As the geometries are so complex, manual methods of calculation have proved difficult to apply and so a combination of manual and computer methods has become common.

1.5 HISTORICAL REVIEW OF OBTAINING STRUCTURAL DESIGN SOLUTIONS BY COMPUTER

1.5.1 The Computer Methods Available

In Chapter 2 the equations of equilibrium (2.5), compatibility (2.12) and stress–strain (2.15) will be developed for problems in linear static elasticity with small deformations. It is the solution of these partial differential equations with appropriate boundary conditions, i.e. loadings and restraints, that provides information that is useful to engineers when carrying out mechanical design. When the problem has simple geometry and simple boundary conditions these partial differential equations can be solved to provide classical solutions. In these cases the solution functions, for example Airy's stress functions in Sec. 2.3.2, are described in the form of series expansions and at every point of the structure they satisfy equilibrium, compatibility and the boundary conditions.

Most engineering designs, however, are too complex for this classical approach. For example, a structure may have spatially dependent material properties if different materials are used; the geometry may be irregular in some sense or the boundary conditions may be complex. In all these examples no solution functions exist and so solutions can be achieved only by resorting to an approximate numerical method.

There are three numerical methods that are commonly used to solve partial differential equations throughout a three-dimensional domain. These methods are:

- the finite element method
- the finite difference method
- the boundary integral or boundary element method.

Of the three methods, the finite element method has become the most widely used when solving structural problems in both industry and academia. There has been a sustained research effort into the finite element method since 1960 and numerous commercial finite element programs are now available. Since then the range of problems being routinely solved has grown, starting with the original application of finding static elastic solutions, to embracing even those problems which possess large deformations and dynamic plasticity.

It is worth a note here to explain some of the philosophy behind the finite element method. Essentially, any problem can be split up into any number of smaller problems. With the finite element method this is done by considering that a complex geometrical shape is made up of a number of simpler shapes. For example, a circle might be approximated by a series of triangles in an attempt to calculate the area of the circle. This is known as *spatial discretization,* with each simple shape being known as an *element* and the whole collection of elements being known as a *mesh.* Within each element the relevant property of the element is predicted, say its area in the case of the triangles approximating a circle, or the relationship between forces and displacements for a structural element. This is done without any reference to other elements in the mesh. Here, people talk of forming the *element equations,* often by assuming known values of properties at fixed points on the elements known as *nodes.* Then the properties of all the elements and the interactions between them are taken into account by *assembling* the element equations and finding a solution to them. In the case of the area of a circle, the element equations calculate the area of the triangles and the solution process adds these together to predict the area of the triangle. Clearly, as more triangles are considered, the area predicted approximates ever more closely the area of the circle, which is known as the *convergence* of the solution.

While the interested reader will find many texts, conferences proceedings, journal papers and review articles concerned with applications of the finite difference method and boundary element method to structural problems, this book focuses on the finite element method. As will be seen from this historical review, developments in mathematics, engineering and computing as shown in Table 1.1 have led to the finite element method's predominant position among the numerical solution methods. Inevitably, the review introduces mathematical terms that may not be familiar to the reader and a better insight into the historical developments may come after reading Chapters 2 and 3. The review is positioned here, to set the scene for the development of the underpinning principles of elasticity and the finite element method.

1.5.2 Some Developments Before 1945

The governing equations of compatibility, equilibrium and stress–strain relationships through material properties were developed well before 1830 by Lamé and Clapeyron. At this time digital computers did not exist and numerical methods for solving partial differential equations were only in their infancy, and so there were very few ways of obtaining useful engineering solutions. As will be discussed in Sec. 2.3, the only way forward for engineers was either to consider simple

Table 1.1 Developments in mathematics, engineering and computing (1830–1980)

Year	Mathematics	Engineering	Computing
1830	Governing equations developed	Use of classical methods	
1840			
1850			
1860			
1870	Calculus of variations		
1880		Use of subregions in structure	
1890		Direct stiffness method for frames	
1900			
1910	Rayleigh–Ritz method		
1920	Galerkin method		
1930			
1940		Use of triangular regions	Digital computer
1950		Use of bars Stiffness method	
1960		Energy methods	
1970		Commerical software appears	
1980			

structures such as bars and beams or to use the stress function approach for a two-dimensional and, on occasions, a three-dimensional continuum.

To progress further, developments in the engineering approach to finding solutions, in the mathematics of solving partial differential equations and in computing technology needed to be made. The first part of the jigsaw was produced by Rayleigh in 1870 when he developed a method for solving partial differential equations based on the minimization of potential energy through the calculus of variations. The only formal difference between this method and the Rayleigh–Ritz method, which was a mathematical stage in the development of the finite element method, is the number of unknowns. In the Rayleigh method the assumed distribution for the unknowns (e.g. displacements of a structure) have a single term only, whereas the Rayleigh–Ritz method can have as many terms as necessary.

Advances soon followed in the handling of the structural problem. In the 1890s engineers moved away from looking at the structure as a single unit, the approach of the method of classical elasticity, where the stress functions are found that satisfy the boundary conditions and also describe exactly the stresses, strains and displacements throughout the whole of the single structure. The new approach was to model the continuum, or structure, as an assembly of simple, structural subregions whose behaviour could be readily described by algebraic equations. Such a spatial discretisation process is natural to engineers and has been used extensively in structural problems ever since.

The most frequently used of these methods has been the *direct stiffness method* associated with structural engineering where buildings and bridges are often designed from slender members. In this area typical problems are analysed as plane frames (skeletal structures), i.e. using one-dimensional beam, torsion and bar members, as will be introduced in Sec. 2.3, and later as finite elements, as will be described in Sec. 3.6.1. Hence it can be seen that, in principle, the method allows complex structures to be analysed providing that the numerical manipulation to solve the algebraic simultaneous equations can be dealt with by hand. Here the development of iterative procedures, such as the moment distribution by Cross (1930) and the relaxation method by Southwell (1940), has benefited the design engineer by enabling solutions to be produced for modest numbers of unknowns without the digital computer. When the computer became available it was logical that engineers would implement the direct stiffness method and this has led to a number of refinements. Traditionally, the connections between members have been assumed to be fully fixed, but recently torsional spring elements have been included to model practical connections in steel frames, known as 'semi-rigid' connections. Refinement of the method allows the analysis to include secondary effects due to changes in the position of the loading as the structure is deformed.

In 1909 Ritz developed the Rayleigh–Ritz method of solving partial differential equations and because of its importance this will be presented in Sec. 3.1.2. This is a generalization of the earlier variational method of Rayleigh. In this method, engineering problems involving the structural continuum can be solved by applying the principle of minimum potential energy to the single body problem, using a simple continuous approximating distribution for the unknowns, which were usually the displacements. Galerkin (1915) showed how such approximate solutions could be derived directly from partial differential equations, leading to the powerful *weighted residuals method* known as the Galerkin method. This is suited to field problems for which the partial differential equations are known but for which a variational statement cannot be found. Examples of such problems are incompressible viscous fluid flow and electromagnetic field theory.

At around the same time, Richardson (1910) used the finite difference method to approximate partial differential equations directly by using the values of the unknowns at points of an imaginary grid to form the derivatives. He used the technique to perform a plane strain analysis of the Aswan Dam. While finite difference procedures have been developed to solve complex problems, they have not become as popular as finite element procedures. There are a number of reasons for this, but in the main the finite difference method is not as good at modelling complex geometries as the finite element method owing to its need for a rectangular grid of points.

It was not until Courant (1943) proposed the idea of using locally defined simple linear distributions defined by specific values at the nodes, in his case for stresses, over a triangular subregion that the forerunner to modern finite elements was developed. His contribution preceded the modern computer and therefore lay dormant until its rediscovery long after the establishment of finite elements.

Clearly, engineers wanted to model a continuum using an assembly of simple structural elements, thereby reducing the problem to a set of simultaneous equations. While the earliest models preceded the modern computer, they were limited to modelling plane frame problems with regular grids of bars or beams. This approach was used by Hrenikoff (1941) and McHenry (1943) who both proposed to approximate the material of an elastic continuum by a lattice of simple bars.

1.5.3 Aircraft Design as a Driving Force (post-1945)

One of the main driving forces in the late 1940s for the development of structural analysis techniques was the need to predict the structural behaviour of aircraft wings. With known materials and simple wing forms the existing analysis methods were adequate, but the increase in aircraft speeds towards and beyond supersonic led to swept and delta wings being proposed. Analysis of these structures was not possible with the existing methods and so new methods had to be developed.

When the stiffness of each subregion is based on an assumed simple stress distribution determined using the forces at the nodes, which are the unknowns to be found, the method is known as the *force matrix method*. However, if the displacements of each node are taken as the unknowns, with an associated simple distribution of displacement assumed throughout the subregion, the method is called the *stiffness method,* and this method was developed by Levy (1953). It was further developed by Turner, Clough, Martin and Topp (1956) who made the major advance of discretizing the continuum fully using an assembly of continuum subregions (or elements) of arbitrary triangular (Fig. 1.2) or quadrilateral form. The domain of the problem, the volume of the physical material itself, was divided into a number of subregions which were considered to be interconnected only at vertices or corners, which are nodes used to define the continuum (Fig. 1.3). Both force and stiffness matrix methods assume a simple piecewise continuous distribution for the unknowns throughout the element. Usually, polynomial functions are used, but occasionally sine and cosine functions are also used. These functions within an element are known within the finite element method by a variety of names such as trial functions or interpolation functions, but here they will be referred to as *shape functions*. Their existence and the restrictions on their form are fundamental to the success of the finite element method and so these will be discussed further in Sec. 3.5.

While there are a number of methods to determine the characteristics of finite elements—the *element stiffness matrix,* [**k**], and the *element force vector,* {**F**^{e}}, in solid mechanics—by 1960, energy methods which assume a simple displacement

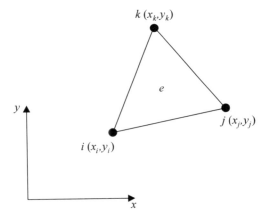

Figure 1.2 A two-dimensional three-noded triangular element.

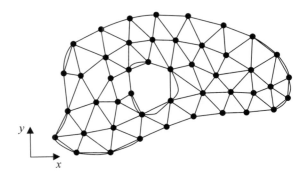

Figure 1.3 An irregularly shaped plate discretized by three-noded triangular elements.

distribution throughout an element had become the standard technique. An important contribution was made by Argyris and Kelsey (1960) who used energy theorems in modelling single triangular and quadrilateral aircraft panels, but the first comprehensive treatment of rectangular finite elements generated by an energy method had been given by Szmelter (1958).

These energy methods came out of a branch of mathematics known as the *calculus of variations,* developed by Rayleigh (1870), and so it can be seen that the wheel has nearly turned full circle, combining mathematics with engineering approaches to produce a workable method for engineering problems in structural design. An explanation and interpretation of this branch of mathematics for the finite element method is outside the scope of this text.

If there is a conservative structural system, i.e. one where the work done by internal forces and the work done by external loads are independent of the path taken, it can be shown that the equilibrium configuration is found by analysis of the potential energy of the system. From the calculus of variations, the variational formulation gives the *principle of minimum potential energy* for small displacement theory of elasticity. In Chapter 3, the finite element representation of this principle will be developed for $[\mathbf{k}]$ and $\{\mathbf{F}^e\}$ and their whole-structure equivalents $[\mathbf{K}]$ and $\{\mathbf{F}\}$. This principle can be regarded as equivalent to the virtual work principle.

So it can be seen that the energy content of a conservative system can be expressed in terms of its configuration without reference to how the system reached its present configuration. Potential or total potential energy includes the strain energy of deformation and the potential possessed by the loading, by virtue of its having the capacity to do work if it moves through a distance. The *principle of minimum potential energy (principle of stationary potential energy)* states that

> Among all admissible configurations of a conservative system, those that satisfy the equations of equilibrium make the potential energy stationary with respect to small admissible variations of displacement.

It is this property of the potential energy that allows us to determine finite element characteristics. Note that the principle is not limited to linear elastic behaviour, but its extension to nonlinear behaviour will be covered only briefly in Chapter 10. Expressions for the potential energy of a structure are integral expressions, called *functionals.* Such functionals implicitly contain the differential equations that state the problem. Whereas the differential equations of classical elasticity (Chapter 2) state a problem in a *strong form* the integrals state it in a *weak form.* The strong form states conditions that must be met at every point in the structure, whereas the weak form states conditions that must be met only in an average sense. This statement indicates that a loss in accuracy must be expected in a finite element analysis, and that it will not necessarily give identical results to a classical solution, if the latter exists. In Sec. 3.1 the steps taken when using the energy method to obtain a general matrix expression for the $[\mathbf{k}]$ and the $\{\mathbf{F}^e\}$, as defined later in Eq. (1.1), will be discussed. Further details on functionals and their interpretation are to be found in many of the texts referred to in this chapter and Chapters 2 and 3.

1.6 OVERVIEW OF THE MODERN FINITE ELEMENT METHOD

The work of Turner *et al.* (1956) can be considered as the birth of the 'finite element method', and the name was first used by Clough (1960) who saw a model as consisting of a *finite number of elements* (or subregions). Now, however, the modelling approach of the finite element method must be discussed so that the basic mathematical method which determines the finite element characteristics can be produced. Here, finite elements based on simple stress shape functions (i.e. the force method) will not be dealt with, as the most commonly used elements are based on displacement shape functions (i.e the stiffness method).

The basic idea behind the finite element method is to construct a structure from a *finite* number of elements, and, as will be demonstrated, these elements may be one-, two- or three-dimensional. In Sec. 3.6.2, the development of the stiffness matrix [**k**] for the three-noded two-dimensional triangular element, as shown in Fig. 1.2, will be shown. When a two-dimensional structure (plane stress and plane strain are defined in Sec. 2.3.2) is constructed of hundreds or thousands of these non-overlapping straight-sided triangles connected at their vertices, it can be seen that all planar geometries can be accommodated. Figure 1.3 shows how an irregularly shaped plate can be discretized into a number of finite elements. At the vertices of the element in Figure 1.2 there are three nodes (i, j, k). These nodes define the elements that discretize the continuum; forces are transmitted by these nodes from one element to the next. The forces acting at the nodes are uniquely defined by the displacements of these nodes, the distributed loading acting on the element, and its initial strain (e.g. temperature, shrinkage, 'lack of fit'). To find the stresses, strains and displacements within the interior of the element, simple displacement functions (the assumed simple displacement distributions) with nodal displacements as the primary unknowns are used to provide a means of interpolation.

For those readers familiar with the direct stiffness method, recall that the behaviour of an element is defined by the matrix equation

$$\{\mathbf{F^e}\} = [\mathbf{k}]\{\boldsymbol{\delta^e}\} \qquad (1.1)$$

in which $\{\mathbf{F^e}\}$ is the *nodal force vector* and $\{\boldsymbol{\delta^e}\}$ is the *nodal displacement vector* containing the element degrees of freedom. The symmetric, positive definite, matrix [**k**] is the *element stiffness matrix,* whose physical meaning is contained in the following statement:

> the *j*th column of [**k**] is the vector of loads that must be applied to nodal displacements in order to maintain the deformation state associated with unit values of displacement *j* while all other nodal displacements are zero.

For the three-noded triangular element in Fig. 1.2, the nodal displacements can be considered to be made up of two independent displacements. Similarly, two independent forces can be assumed to act at each node. These independent quantities are usually defined in terms of a Cartesian coordinate system, i.e. both forces and displacements have components in the *x*- and *y*-directions, such that at node *i* the nodal displacements are u_i and v_i and the nodal forces are F_{xi} and F_{yi}. By the same reasoning, the element vectors $\{\mathbf{F^e}\}$ and $\{\mathbf{\delta^e}\}$ have six terms and thus $[\mathbf{k}]$ is a 6 by 6 matrix. Each nodal displacement in $\{\mathbf{\delta^e}\}$ is one of the degrees of freedom for the element and so our triangular element has a total of 6. It is the formulation of the terms in the finite element characteristics $[\mathbf{k}]$ and $\{\mathbf{F^e}\}$ that is fundamental to the finite element method, and Sec. 3.4 discusses in detail the methods available.

There are two principal reasons why the method has become the preferred numerical method to solve linear static structural problems using many discrete points (i.e. nodes) in modelling, and these are:

1. Once the stiffness and nodal force vector matrices in terms of geometry, nodal coordinates and material properties for an element type are known, the calculation of their terms for each occurrence of that element in the model is straightforward.
2. It is a simple process to take the individual element stiffness matrix and nodal force vector and then construct and solve the algebraic equations for the whole structure.

In matrix notation the equation representing the whole structural problem is

$$\{\mathbf{F}\} = [\mathbf{K}]\{\mathbf{\delta}\} \tag{1.2}$$

in which both the *vector of nodal force*, $\{\mathbf{F}\}$, and the *vector of nodal displacements*, $\{\mathbf{\delta}\}$, have n terms. $[\mathbf{K}]$ is the *stiffness matrix* (sometimes prefixed by *global* or *structure*) having n by n terms. It is not uncommon for models of commercial problems to have the number of degrees of freedom n in the order of tens of thousands, these being the combination of all the individual element degrees of freedom (each of which provides components in their $\{\mathbf{\delta^e}\}$).

The procedures used in commercial software to generate the element characteristics and algebraic equations, to apply boundary conditions to them, and to solve for nodal displacements and element stress will be given later in Secs 3.6–3.9.

1.7 INDUSTRIAL IMPLEMENTATION AND RESEARCH IN FINITE ELEMENTS

Having given an engineering insight into the historical development of the finite element method, it is now time to discuss some of the other advances since 1870 that have contributed to making the method so successful in industry and academia. From the above discussion it can be seen that any structural problem can, in principle, be solved using the finite element method if a mathematical method is available to derive the element stiffness and element force matrices for a structural problem under consideration. From our knowledge of classical elasticity, as will be outlined in Chapter 2, it seems sensible to apply this method to the calculation of finite element equations. A classical solution requires its functions (e.g. Airy's stress functions) to satisfy, at the same time, not only the differential equations but also both displacement (or *essential*) and the stress (*non-essential* or *natural*) boundary conditions. Consequently, solutions are found only for a complete structure and are limited to structures with simple geometries and boundary conditions. Such a restriction means that classical elasticity cannot be used to find the [**k**] for a small subregion in the interior of a structure.

An important method for deriving element stiffness matrices [**k**] is the direct method that is based on physical reasoning. It is limited to simple linear elastic elements such as a bar and a pure bending beam with constant properties along their lengths. Note that such elements are identical to those used in the direct stiffness matrix method. Attention to this simple method is often found in texts because it enhances the reader's understanding, but real structures cannot often be modelled using such simple elements so alternative methods are sought to develop [**k**] when shape, material properties and behaviour are more complex.

Both variational and weighted residual methods have been developed to derive [**k**] using piecewise continuous distributions for the nodal unknowns. For structural mechanics problems it is observed that the finite element characteristics are identical whichever of the two methods is used.

An interesting revelation to come out of the historical development before 1960 is that many methods, in different branches of mathematics, were independently developed, and that these were later brought together in the finite element method as the digital computer made its impact in permitting a direct solution of quite large systems of algebraic equations. As the 1960s progressed major advances took place in development and research, and these were accompanied by a growth in the number of publications from 10 in 1961 to 531 in 1969. The first textbook on the subject was written in 1967 by Zienkiewicz and Cheung. An exponential growth in research, development and publications continued in the 1970s. Specific journals for papers on numerical methods in engineering were created and their number and size have continued to grow ever since. Commercial finite element programs appeared in the market-place in the 1970s and the pace at which the finite element method has been allowed to develop mirrors the increasing capacity of suitable computing hardware up to the current situation described in Chapter 4.

The status today is that the annual contribution of papers continues to grow and it is estimated that some 5000 papers are published each year. Also, while mathematicians continue to put the method on firm theoretical ground, engineers continue to find new applications and extensions in many fields. The range of problems solved by structural finite element analyses has become extensive such that commercial programs can, for example, analyse successfully the behaviour of highly nonlinear plastic structures under impact. NAFEMS (National Agency for Finite Element Methods and Standards) in the United Kingdom has estimated that some of the major companies writing finite element software now have an annual turnover of some £50m each. From this it is estimated that the total annual investment is around £1 billion.

Future research will, no doubt, concentrate on tools to aid the design engineer such as error estimation and adaptive analysis. The number of algebraic equations to model a problem will also get larger and this will concentrate development on stable iterative procedures, the use of parallel processors and a return to model-ling with simple elements.

Recently there has been a development in the monitoring and controlling of programs. Users of finite element packages may be so impressed by the versatility of the pre- and post-processing features available that modelling and analysis limitations are ignored. As many of the commercial programs do not have accu-racy estimation built-in, so users must learn by experience to produce accurate solutions. It can be reasonably assumed that all packages contain bugs either in terms of implementing wrong modelling assumptions or by introducing incorrect coding. To help users with problems when solving in real commercial situations, organizations such as NAFEMS and similar bodies in the United States have been formed. The publication of benchmark examples by these authorities for a range of problems has enabled users to test the accuracy of their own modelling technique when using software.

1.8 REFERENCES AND FURTHER READING

Interested readers might wish to look at some key texts. For example, the design process is discussed by Deutschman, Mitchels and Wilson (1975), Shigley and Mitchell (1983) and Pahl and Beitz (1988), and Lamit (1994) gives a good account of the design variables that need to be considered. Tizzard (1994) is an up-to-date text that discusses concurrent and simultaneous engineering as well as the skills needed by engineers. Similarly, Onwubiko (1989) looks at the benefits of computer-aided design.

Works on basic structural problems include Clapeyron and Lamé (1833), Airy (1863), Maxwell (1872), Rayleigh (1870), Ritz (1909), Galerkin (1915) and Timoshenko (1953). Slender members are discussed by Coates, Coutie and Kong (1988).

Key works on numerical methods include Richardson (1910), Cross (1930), Southwell (1940) and Shaw (1992).

For a review of the development of finite element methods, see Hrenikoff (1941), Courant (1943), McHenry (1943), Levy (1953), Turner *et al.* (1956), Szmelter (1958), Argyris and Kelsey (1960), Clough (1960), Zienkiewicz and Cheung (1967), Martin and Carey (1973), Washizu (1982), Norrie and de Vries (1976), Reddy (1984), Cook, Malkus and Plesha (1989), Zienkiewicz and Taylor (1989), NAFEMS (1992) and Cook (1995).

2

STRUCTURAL MECHANICS

This chapter discusses the ways in which structures behave when they are subjected to loading. At first the development will be intuitive, but then a mathematical development of the description of a structure's behaviour will be given. Here the class of behaviour considered is that of small deformation of structures of a single material that is isotropic and generally linear elastic. Once a general mathematical description is available, the description may be simplified in certain situations and some of these simplifications will be outlined. In particular, it will be shown that manipulation of the general description leads to the simple formulae that engineers are taught during their first courses in structural mechanics. It is the solution of the equations of this same general description that computational methods such as the finite element method have been developed to solve. The chapter concludes with an outline of how problems in structural design are described.

Development of the finite element equations will be covered in Chapter 3 and the presentation of advanced topics such as time-dependent and nonlinear behaviour will be left until Chapter 10.

2.1 THE USE OF LOAD-BEARING STRUCTURES

There are examples of the use of load-bearing structures all around, but do people ever think about what the structures are actually doing? As the answer to this is probably 'no', the behaviour of load-bearing structures must be considered in some detail. Take, for example, a floor supporting a weight such as a person sitting on a chair, or shelves that are used to carry heavy books. In each of these cases the structure acts to hold items, and itself, in place against the effects of gravity loading.

All objects have a mass, and an object's weight is the effect of gravity on the mass. This weight is a body force (i.e. internal load) and it acts on and is resisted by the structures mentioned above. If the structure can provide an equal and opposite force to that due to gravity, then the total force on the object will be zero and, if it is at rest, the object will remain at rest and stay in position.

Commonly, there are two parts to the load. A structure's self-weight is known as the *dead load*. For all but massive structures (e.g. bridges and aerospace vehicles) the dead load is small compared to that part of the load which is applied externally to the structure. This latter load is known as the *live load*. In the case of the books on the bookshelf the load's cause is the effect of gravity. In other circumstances, however, the live load can be due to other effects such as wind, snow, impact, acceleration and friction. For example, in machines many forces are generated by the operation of the machine. Consider a bicycle, where power from the rider is used to propel the bike–rider combination. Gravity acts on the rider and the bike, and so the wheels have to resist a gravitational dead load acting on the ground. However, the rider also exerts a resultant force on the pedals to provide the motive power, and so the pedal axle must resist this force and its associated moment which in turn has to be resisted by the crank and so on. Eventually the force exerted by the rider must be transmitted through the pedal, crank, gears and chain to the rear wheel. Hence, it can be seen that structures must not only resist forces but they must transmit them as well.

Let us consider an elastic band. If the band is pulled it resists the force that is exerted, and the greater the pull the greater the resistance. Structures may be thought of as being like a whole series of elastic bands or even springs, where forces between the atoms or molecules of the structure resist the deflection of the structure due to the external forces acting on it. When a bar is pulled, it is under tension, and if a bar is pushed it is placed under compression. A beam can also be bent by applying a moment to the beam that acts perpendicularly to its longitudinal axis, or a circular bar can be placed under pure torsion by twisting it, i.e. exerting a moment that acts in the plane perpendicular to the longitudinal axis of the bar. However, no matter which way it is loaded, be it under tension, compression, bending or torsion, the force exerted on the structure is resisted by the internal forces generated by moving the atoms or molecules of the material of the structure.

With an elastic band the deflection is large, but often structures are not thought of as deflecting under load. However, let us consider a shelf full of books, where it is quite often noticeable that the shelf deflects several millimetres because of weight of the books. This deflection is the result of the atoms or molecules in the shelf being moved apart or moved together. When this happens the interatomic forces that are generated provide the necessary force to oppose the weight of the books, and clearly the deflection depends on the properties of the structure's material.

From these simple examples the key ideas underlying the analysis of structures can be listed. If a structure is considered that has external loading applied to it, and the structure neither moves as a rigid body after loading nor fails, i.e. it resists load by deflecting only a small amount, then it is known that:

1. The structure must be in *equilibrium*, with the loading being matched by appropriate reactions where the structure is restrained, such that the total force and moment acting on the structure are zero. The structure is therefore either at rest or travelling with a constant velocity.
2. The structure will deflect a small amount as the atoms or molecules of the material move and so generate forces inside the structure. Hence the structure is *stressed* internally, with each piece of material in local equilibrium and subject to *strains* which are a measure of the deformation of the structure and will be defined later.
3. The topology of the structure remains the same even though it has deflected. That is, the material of the structure moves such that the form remains the same, and it is said that *compatibility* of the structure is maintained.
4. The size of the deflections is determined by the *material elastic constant properties* of the structure.

Now, having studied the physical nature of structural mechanics the mathematical rules that govern the behaviour can be determined.

2.2 ENGINEERING STRESS AND STRAIN

Having established that all structures deform when subjected to a set of external and restraining forces that are in equilibrium, and that these forces are transmitted through the body, quantities are needed to describe the internal response, or state, of the structure. These quantities can be defined by considering the concepts of engineering stress and strain. The analysis of stress is essentially a branch of statics that is concerned with the detailed description of the way in which the stress (or intensity of force) at a point in the structure varies, whereas the analysis of strain is essentially a branch of geometry which deals with the deformation of an assemblage of particles. The variation of stress and strain throughout a structure, unless it is fracturing, is found to be continuous and described by single-valued functions. These functions cannot be found readily except for cases where the geometries and loading are simple, as discussed in Sec. 2.3.

2.2.1 Stress at a Point

The understanding of stress at a point is fundamental to stress analysis. In order to specify the forces acting within a structure an arbitrary plane can be chosen which passes through the material. On this plane the point O is located at the centroid of a small flat surface of area δA, as shown in Fig. 2.1(a). The *positive side* of the plane is defined by the outward normal to the plane ON, as shown, and the plane in the opposite direction is the *negative side*.

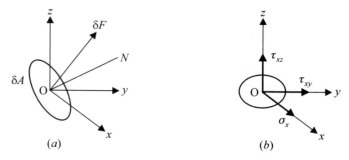

Figure 2.1 The forces and stresses on a small area: (*a*) forces, (*b*) stresses.

At all particles on the surface of δA the material on one side of the surface exerts a force upon the material on the other side, so that equilibrium is maintained if a cut is made along our arbitrary plane and these forces are applied at the cut. The resultant of the forces exerted by the material on the positive side of δA is a force δF. A force δF also acts on the material on the negative side of δA in the opposite direction to maintain equilibrium. Note that the resultant force is assumed to have its line of action through point O and to have produced no couple. Now, the total stress at the point O across the plane whose normal is in the direction ON is defined by

$$\text{Total stress}_{ON} = \lim_{\delta A \to 0} \frac{\delta F}{\delta A} \tag{2.1}$$

a vector whose magnitude has dimensions of force per unit area. In the physical situation, for every point O of the body and every plane through each point, a vector of total stress exists. Hence it can be seen that, to be strictly accurate, stress is not a vector quantity as not only the magnitude and direction of the vector but also the plane on which the stress acts must be defined. Stress is therefore a tensor, its complete description depending on the vectors of force (δF) and the outward normal to the surface of action (a vector in the direction ON).

2.2.2 Types of Stresses

It is convenient when developing the mathematical theory of stress analysis to set up a rectilinear coordinate system, such as the Cartesian, polar or spherical coordinate systems. Here the Cartesian system will be used as this is the most common in classical elasticity and the finite element method. Now place point O at the origin of three mutually perpendicular right-handed axes Ox, Oy and Oz. Let the normal ON be in the direction Ox as in Fig. 2.1(b). The surface δA lies in the y–z plane. The total stress can be resolved into three components σ_x, τ_{xy} and τ_{xz} in directions Ox, Oy and Oz respectively. The component σ_x, which acts in the direction normal to the plane, is called the *direct* or *normal stress*. The other two components τ_{xy}, and τ_{xz}, which act in the plane, are called *shear stresses*.

Two subscripts are needed to define a shear stress; the first subscript denotes the direction of the normal to the small area δA and the second subscript the direction in which the shear stress acts. Only one subscript is needed to define the direct stress since the direction is the same as that of the normal to the surface. This sign convention is related to the deformational influence of the stress. If σ is positive (Fig. 2.1b) it is called a tensile stress (i.e. the stress tends to stretch the material on the positive side of the surface away from that on the negative side), and if it is negative it is called a compressive stress.

The above argument can be extended to establish the three-dimensional stresses at a point. Figure 2.2 represents an infinitesimal rectangular parallelepiped (cuboid) with its sides parallel to a Cartesian coordinate system. The surface δA is now each of the six sides of the parallelepiped. The length of each side is taken to be so small that the stress components are uniformly distributed on each face and that they do not vary from the negative to the positive side. A three-dimensional state of stress at a point is shown in Fig. 2.2, where the directions of the stresses shown are defined to be positive.

The centroid of the parallelepiped is at C. A total of nine stress components define the state of stress at point C. By taking moments and considering equilibrium, it can be shown that the shear stresses have to be complementary. Hence

$$\tau_{xy} = \tau_{yx}$$
$$\tau_{yz} = \tau_{zy} \qquad\qquad (2.2)$$
$$\tau_{zx} = \tau_{xz}$$

The six quantities σ_x, σ_y, σ_z τ_{xy}, τ_{yz} and τ_{zx} are therefore both sufficient and necessary to define the stresses acting on the coordinate planes through point C.

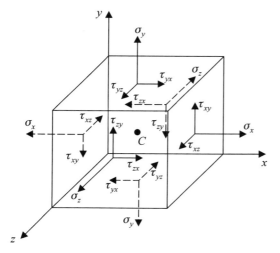

Figure 2.2 The stresses on an infinitesimal volume.

The parallelepiped in Fig. 2.2 can be rotated to any orientation within the structure about its centroid C. At each orientation the stresses, both direct and shear, acting on its faces are in static equilibrium. This gives mathematical relationships for these stresses in terms of the angles of rotations and stress components referenced to the Cartesian coordinate system. A visual representation for the variation of the stress components on the faces as the parallelepiped is rotated is obtained using Mohr's circle for three-dimensional stress. At one specific orientation it is found that there are no shear stress components on any of the faces. Then the three direct stress components that exist are known as the principal stresses (Fig. 2.3). These are given the notation of σ_1, σ_2, and σ_3, where the subscripts define the axes of the principal coordinate system. Their directions are along the principal axes, and the planes on which they act are the principal planes. The maximum shearing stress acts on one of the planes at 45° to the planes on which the principal stresses act, and its value is given by

$$\tau_{max} = \frac{\sigma_{max} - \sigma_{min}}{2} \tag{2.3}$$

where the direct stresses are the maximum and minimum principal stresses (i.e. the maximum is the most tensile, and the minimum is the most compressive). For example, in Fig. 2.3 $\tau_{max} = \tau_2$ when $\sigma_1 > \sigma_2 > \sigma_3$. It is important to understand that principal stresses are readily derived by transformation from Cartesian values and that they are the parameters in conventional failure criteria (Sec. 2.2.6) that are often the useful output of a finite element analysis.

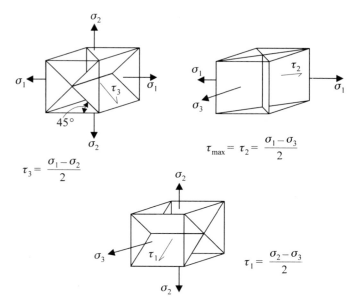

Figure 2.3 The principal stresses.

2.2.3 Differential Equations of Equilibrium

If the lengths of the sides of the infinitesimal parallelepiped in Fig. 2.2 are allowed to increase, it is now not a point that is considered but a small subregion enclosing a certain volume of material. In the general case, the direct and shear stresses on opposite faces are not equal. This variation in stresses is indicated in Fig. 2.4. Note that the shear stresses are defined using the notation on the left-hand side of Eq. (2.2). Thus if the direct stress on the negative-facing plane of constant x is σ_x then the direct stress acting on the positive-facing plane at $x + \delta x$ is, from the first two terms of a Taylor series expansion, $\sigma_x + (\partial \sigma_x / \partial x)\delta x$. It is assumed that the stresses acting on each face of the subregion do not vary over the surface such that their lines of action go through the centroid, as shown in the Fig. 2.4.

To establish a relationship for equilibrium the forces acting on each face of the subregion must be considered. It is the six stress components that produce these surface forces. For example, the resultant force in the x-direction on the negative-facing plane of constant x is $\sigma_x \delta y \delta z$. In the process of increasing the size of the rectangular parallelepiped from a point to a subregion material is introduced into the analysis of volume $\delta x \delta y \delta z$. This provides forces distributed over the volume of the subregion, such as gravitational forces, magnetic forces, or in the case of motion, inertia forces, known as *body forces*. The body forces per unit volume are resolved into the three Cartesian directions and the components are denoted by X, Y, and Z.

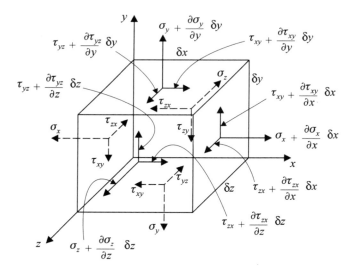

Figure 2.4 The stresses on the faces of a volume element.

In Fig. 2.4 the subregion is in equilibrium, acted upon by the surface forces and the components of the body forces, which are not shown. Considering equilibrium in the x-direction

$$\left(\sigma_x + \frac{\partial \sigma_x}{\partial x} \delta x\right)\delta y \delta z - \sigma_x \delta y \delta z + \left(\tau_{xy} + \frac{\partial \tau_{xy}}{\partial y} \delta y\right)\delta x \delta z$$

$$-\tau_{xy}\delta x \delta z + \left(\tau_{zx} + \frac{\partial \tau_{zx}}{\partial z} \delta z\right)\delta x \delta y - \tau_{zx}\delta x \delta y + X\delta x \delta y \delta z = 0 \qquad (2.4)$$

If similar expressions are written for the y- and z-directions, and then simplified, then

$$X + \frac{\partial \sigma_x}{\partial x} + \frac{\partial \tau_{xy}}{\partial y} + \frac{\partial \tau_{zx}}{\partial z} = 0$$

$$Y + \frac{\partial \tau_{xy}}{\partial x} + \frac{\partial \sigma_y}{\partial y} + \frac{\partial \tau_{yz}}{\partial z} = 0 \qquad (2.5)$$

$$Z + \frac{\partial \tau_{zx}}{\partial x} + \frac{\partial \tau_{yz}}{\partial y} + \frac{\partial \sigma_z}{\partial z} = 0$$

If no body forces exist, then X, Y, and Z are all zero. This is often the situation in mechanical design as the weight of a structure is much less than the forces it must transmit, but note that the body forces have to be included in the analysis when their magnitudes are of similar order to those of the external forces.

2.2.4   Strain at a Point

The stresses described in Sec. 2.2.2 cause small linear and angular displacements in the deformable structure. These small displacements are generally defined in terms of engineering strains, which are also known as Lagrangian strains. Note that there are a number of definitions for strain and each has its appropriate mathematical foundations. Engineering direct strains define the change in length produced by direct stresses while engineering shear strains define the change in angle produced by shear stresses. These strains are denoted, with appropriate subscripts, by ε and γ respectively, and they have the same sign convention as the associated stresses (i.e. σ_x and ε_x, τ_{xy} and γ_{xy}, etc.).

The displacement of any portion of a loaded structure can first be resolved into components u, v and w parallel to the x-, y- and z-axes, and the functions that describe the variation of these displacements over the volume of the structure are continuous and singled-valued. Hence, the displacement in the x-direction, u, is defined by a function of x, y and z. It will be assumed that these components of displacement are very small quantities, and so, as was seen when the stress at a point was analysed in Sec. 2.2.2 an approximation for the strain can be made. Provided that the subregion of material being considered is sufficiently small, the strain within the material may be treated as a constant or homogeneous, giving the definition in the limit to the concept of strain at a point.

Consider the very small subregion with sides of length δx, δy and δz shown in Fig. 2.5(a) which is part of the undeformed structure. Choose three mutually perpendicular lines PA, PB and PC at a point P. If the structure undergoes a small deformation with displacement components u, v, and w at point P, it will move to P', as shown in Fig. 2.5(b). Similarly, A goes to A', B to B' and so on. The displacement in the x-direction of the point A on the x-axis is approximately $u + (\partial u / \partial x)\delta x$ (i.e. the first two terms of a Taylor series expansion). The

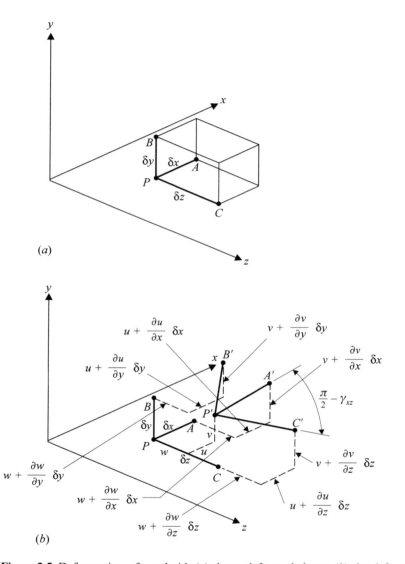

Figure 2.5 Deformation of a cuboid: (a) the undeformed shape, (b) the deformed shape.

remaining displacement components of the point A and those of the points B and C can also be found and are shown in Fig. 2.5(b).

Note that the approximation, in which second- and higher-order terms of the Taylor expansion are neglected, is consistent with the assumption of constant strains (i.e. constant displacement gradient) over the subregion. It is also consistent with the straight line PA in the undeformed structure becoming the displaced straight line $P'A'$ following deformation. The second-order terms would be included in the development of a geometric nonlinear analysis.

If a line of length L_o undergoes a change in length ΔL then direct strain is defined as

$$\varepsilon = \frac{\Delta}{L_o} = \frac{L - L_o}{L_o} \tag{2.6}$$

where L is the final length of the line. This is known as the *engineering direct strain* and is acceptable for strains of magnitude less than 0.1. Note that the *true* strain is required when the extension of the current length is considerable. For true strain the change in length is referred to the instantaneous length, rather than to the original length.

The change in length of the line PA is $P'A' - PA$ so that the direct strain at P in the x-direction is obtained from

$$\varepsilon = \frac{P'A' - PA}{PA} = \frac{P'A' - \delta x}{\delta x} \tag{2.7}$$

From considerations of geometry and applying the binomial expansion

$$P'A' = \delta x \left(1 + \frac{\partial u}{\partial x} \right) \tag{2.8}$$

in which second- and higher-order terms of $\partial u / \partial x$ are neglected. Substituting for length $P'A'$ in (2.7)

$$\left. \begin{array}{c} \varepsilon_x = \dfrac{\partial u}{\partial x} \\[2mm] \text{and similarly} \qquad\qquad \varepsilon_y = \dfrac{\partial v}{\partial y} \\[2mm] \varepsilon_z = \dfrac{\partial w}{\partial z} \end{array} \right\} \tag{2.9a}$$

Engineering shear strain at a point is defined as the change in the angle between two mutually perpendicular lines intersecting at the point. Therefore, if the shear strain in the x–z plane is γ_{xz} then the angle between the displaced lines $P'A'$ and $P'C'$ in Fig. 2.5(b) is $\pi/2 - \gamma_{xz}$ radians. From the geometry and displacements in Fig. 2.5(b) it can be shown that

$$\gamma_{xz} = \frac{\partial u}{\partial z} + \frac{\partial w}{\partial x}$$

$$\gamma_{yz} = \frac{\partial v}{\partial z} + \frac{\partial w}{\partial y} \qquad (2.9b)$$

$$\gamma_{xy} = \frac{\partial u}{\partial y} + \frac{\partial v}{\partial x}$$

The shear strains are complementary and so $\gamma_{xy} = \gamma_{yx}$, $\gamma_{yz} = \gamma_{zy}$, $\gamma_{zx} = \gamma_{xz}$. As strains are tensor quantities, like stresses, it follows that there are three principal strains (i.e. ε_1, ε_2 and ε_3) whose magnitudes and directions can be obtained by transformation. This gives Mohr's circle for strain. It is worth noting that when the material has the same mechanical properties in all directions, i.e. it is isotropic, then the principal stresses and principal strains are aligned.

It is convenient when formulating the finite element stiffness matrix (Chapter 3) to use matrix notation and so the strain–displacement relationships (2.9a,b) can be rewritten as $\{\varepsilon\} = [\partial]\{u\}$ or in full

$$
\begin{Bmatrix} \varepsilon_x \\ \varepsilon_y \\ \varepsilon_z \\ \gamma_{xy} \\ \gamma_{yz} \\ \gamma_{zx} \end{Bmatrix}
=
\begin{bmatrix}
\frac{\partial}{\partial x} & 0 & 0 \\[4pt]
0 & \frac{\partial}{\partial y} & 0 \\[4pt]
0 & 0 & \frac{\partial}{\partial z} \\[4pt]
\frac{\partial}{\partial y} & \frac{\partial}{\partial x} & 0 \\[4pt]
0 & \frac{\partial}{\partial z} & \frac{\partial}{\partial y} \\[4pt]
\frac{\partial}{\partial z} & 0 & \frac{\partial}{\partial x}
\end{bmatrix}
\begin{Bmatrix} u \\ v \\ w \end{Bmatrix}
\qquad (2.10)
$$

2.2.5   Compatibility Equations

Equations (2.9), or (2.10), express the six components of strain at a point in a structure in terms of the three components of displacement u, v and w at that point. For the strains to be compatible over the volume then the material cannot overlap or form gaps (i.e. no voids are created). It follows then that the components of displacement can be given by

$$u = f_1(x, y, z)$$
$$v = f_2(x, y, z) \qquad (2.11)$$
$$w = f_3(x, y, z)$$

where the functions are single-valued and continuous. If gaps are produced due to deformation, the displacements are discontinuous and separate functions either side of a gap are necessary. The existence of just three singled-valued functions for displacement is a statement of the continuity or compatibility of displacement.

Since the six strains described by (2.9) are functions of only three displacements, they cannot be independent. By differentiating these equations twice and substituting, the following six compatibility equations are derived:

$$\frac{\partial^2 \gamma_{xy}}{\partial x \partial y} = \frac{\partial^2 \varepsilon_x}{\partial y^2} + \frac{\partial^2 \varepsilon_y}{\partial x^2}$$

$$\frac{\partial^2 \gamma_{yz}}{\partial y \partial z} = \frac{\partial^2 \varepsilon_y}{\partial z^2} + \frac{\partial^2 \varepsilon_z}{\partial y^2}$$

$$\frac{\partial^2 \gamma_{zx}}{\partial x \partial z} = \frac{\partial^2 \varepsilon_z}{\partial x^2} + \frac{\partial^2 \varepsilon_x}{\partial z^2}$$

$$2\frac{\partial^2 \varepsilon_x}{\partial y \partial z} = \frac{\partial}{\partial x}\left(-\frac{\partial \gamma_{yz}}{\partial x} + \frac{\partial \gamma_{zx}}{\partial y} + \frac{\partial \gamma_{xy}}{\partial z}\right)$$

$$2\frac{\partial^2 \varepsilon_y}{\partial x \partial z} = \frac{\partial}{\partial y}\left(\frac{\partial \gamma_{yz}}{\partial x} - \frac{\partial \gamma_{zx}}{\partial y} + \frac{\partial \gamma_{xy}}{\partial z}\right)$$

$$2\frac{\partial^2 \varepsilon_z}{\partial x \partial y} = \frac{\partial}{\partial z}\left(\frac{\partial \gamma_{yz}}{\partial x} + \frac{\partial \gamma_{zx}}{\partial y} - \frac{\partial \gamma_{xy}}{\partial z}\right)$$

$$(2.12)$$

To gain further insight into the principle of compatibility, imagine a structure subdivided into a large number of small rectangular parallelepiped subregions before deformation. On loading the structure these subregions will be deformed into a system of general parallelelepipeds. While the subregions are now deformed, they must be connected to each other in exactly the same way as before, i.e. the topology of the structure remains the same. If the components of strain satisfy the compatibility equations, then this situation will be forced to occur, and therefore the six equations of (2.12) must be satisfied when solving any problems in small-displacement elasticity.

2.2.6   Structural Materials and Hooke's Law

One consideration of the functional design of mechanical structures is that the materials used have adequate strength and stiffness. These materials are referred to as structural materials (e.g. steels, aluminium and titanium alloys, engineering composites, timbers and plastics). To carry out a successful stress analysis for the purpose of design analysts must have a knowledge of the material properties, in particular the elastic constants (e.g. Young's modulus and Poisson's ratio) and strengths. Other properties such as thermal conductivity, wear resistance and corrosion resistance may be relevant to the product's function, but are not of concern in this treatment of the fundamental principles.

Throughout most of the text, an important assumption is made that a structure consists of material which is continuous, homogeneous and isotropic. A body is homogeneous if it has identical properties at all points and it is considered to be isotropic when its properties do not vary with direction or orientation. Except for Chapter 10 the development of stress analysis will be limited to materials that are linear elastic. While a number of traditional structural materials such as steels, aluminium alloys, particulate composites (such as concrete) and some plastics appear to meet these conditions when viewed on the macroscopic scale, it is apparent when they are viewed on the microscopic scale that they are anything but homogeneous and isotropic. For example, metals are generally made up of more than one phase, in the form of small crystal grains with different properties (strengths and moduli), such that they are actually heterogeneous. The reason why the stress analysis equations given here are still valid for the behaviour of metal structures is that the very large number of heterogeneous grains are uniformly distributed over the volume; thus making the polycrystalline material statistically homogeneous and isotropic. As a rule of thumb, a structure satisfies the requirement if its dimensions are at least 20 times the largest dimension of any grain or inclusion. Since most structures are considerably larger than the microstructural detail it is acceptable to use the assumptions stated.

A property which varies with orientation is said to be an anisotropic property. Metals become anisotropic when they are deformed severely in a particular direction, as happens in rolling and forging. Other examples of structural materials with anisotropic properties are fibre-reinforced composites, reinforced concrete, oriented plastics and timbers.

Having looked at the assumptions, let us return to the mathematical development. So far nine equations (2.5) and (2.9), or (2.5) and (2.12), have been produced for a deformable structure, involving fifteen unknowns (six stresses, six strains and three displacements). For a solution further equations must be found. These equations are provided by six material relationships which describe the stress–strain behaviour of the material. Note that the derivation of (2.5) and (2.9) does not make any assumption about this behaviour. It follows, then, that these fundamental equations are applicable to any type of deformable structure no matter how complex the material behaviour is.

The procedure for establishing the stress–strain behaviour utilizes certain simple experiments (e.g. in a standard testing machine) in which both stress and strain are measured and the relationship between them determined. This approach is used to produce idealized mathematical formulae relating stress and strain from which the response of any structures under more complicated deformation can be calculated. Note that the properties of an engineering material vary enormously with temperature, pressure, rate of strain and fatigue, so that any formula is only a reasonable approximation to the response of the material under limited conditions. For now, however, consider only the case of a material under normal laboratory conditions, which is an example of the ideal situation, being simple and one-dimensional.

Suppose that a bar of constant cross-sectional area is stretched in tension. The rate of stretching is such that fracture occurs within several minutes. Measurements from the test are therefore not affected either by a high strain rate or by the length of time the load is applied. Such a tension test is used to measure the tensile elastic constants of metals, plastics, fibre reinforced composites and so on, and is conducted in accordance with international and national standards. It is often known as a *coupon test*. The measured quantities are engineering direct stress σ_a (where subscript 'a' denotes pure axial loading), which is the load applied divided by the original cross-sectional area and engineering direct strain ε_a which is $(L - L_o)/L_o$ (Eq. 2.6), where L_o is the original length and L is the length when the stress is σ_a. If σ_a is plotted against ε_a, stress–strain curves such as those shown in Fig. 2.6 are obtained. Note that the area of the bar changes during the test, so that strictly σ_a as defined is not the true

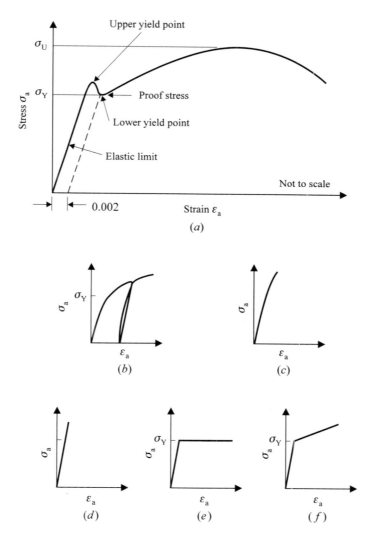

Figure 2.6 Some stress–strain curves for structural materials: (*a*) ductile steel, (*b*) aluminium alloy, (*c*) unidirectional fibre-reinforced composite material, (*d*) linear elastic model, (*e*) perfect linear elastic-plastic model, (*f*) linear elastic-strain hardening model.

stress. However, for small strains of magnitude less than 0.02 (i.e. ±2 per cent) the difference between the stresses is small enough to be ignored, and the theory of infinitesimal strain described in Sec. 2.2.4 remains acceptable.

It is found that for an applied stress below a certain limit, if the stress is reduced to zero the bar returns to its original dimensions, i.e. there is no permanent deformation. This is the property of the material known as *elasticity*, and the range of stresses in which there is no permanent deformation is the *elastic range*. The most widely used structural materials, such as metals, composites and timbers, have an initial elastic stress–strain response which for mathematical modelling purposes can be assumed to be linear. This provides Fig. 2.6(d), the idealization of the perfectly linear elastic material. It is of paramount importance in mechanical design because many structures are designed to operate with low values of stress within them so that a linear stress analysis and thus the principle of superposition will be valid.

The linear relationship between stress and strain for a bar in tension (or compression) gives the definition of the modulus of elasticity, E, which is also known as Young's modulus, from

$$\sigma_a = E \varepsilon_a \qquad (2.13)$$

Hence E is defined by the slope of the stress–strain curve in the linear elastic region, and its value depends upon the particular structural material being used. For materials that do not possess a linear elastic stress–strain range the definition for E is changed such that (2.13) may still be used. The quoted E for plastics is the *secant* modulus of elasticity. Table 2.1 presents values for a variety of material properties including Young's modulus for a number of structural materials. Unlike strength, it is insensitive to the precise microstructure. For example, steels have a value for E of 205–215 kN mm^{-2} (GPa) while the yield strength can range from 250 to 1900 N mm^{-2} (MPa). Equation (2.13) is commonly known as Hooke's law, but it is a very limited version of the generalized Hooke's law as it relates only the direct stress and strain developed in a bar subjected to pure axial loading.

As the bar undergoes longitudinal stretching there is an accompanying lateral contraction. Equally, lateral expansion is observed if the bar is compressed. In the linear elastic range, it is found experimentally that the lateral direct strains, say in the y- and z-directions, are related to the longitudinal direct strain, in the x-direction, by a constant, ν, such that

$$\varepsilon_y = \varepsilon_z = -\nu \varepsilon_x = -\frac{\nu \sigma_x}{E} \qquad (2.14)$$

where ν is called Poisson's ratio (see Table 2.1).

For the three-dimensional state of stress, each of the six stress components is expressed as a linear function of six components of strain. This is the generalization of Hooke's law and can be written as

Table 2.1 Typical material properties of structural materials

Material	Young's modulus E kN mm^{-2} (GPa)	Poisson's ratio ν	Yield stress (or yield strength) σ_Y N mm^{-2} (MPa)	Ultimate strength[1] σ_U N mm^{-2} (MPa)
Carbon steels	205	0.30	250–350	400–700
1.5% Mn steels	205	0.3	320–500	540–850
Stainless steels	215	0.3	280–1900	520–1970
Cast iron (malleable)	170	0.25		270–570
Aluminium cast alloys	71	0.30	55–250	145–330
Concrete	15–40	0.10		15–70[2]
Timbers[3]	3.5–17			20–130
Plastics	0.06–11			5–130
Unidirectional fibre reinforced polymers[4]	35→350	0.25–0.35		400→2000

Notes

1. Tensile strength unless otherwise stated
2. Compression strength
3. Parallel to grain
4. Parallel to fibres

$$\sigma_x = C_{11}\varepsilon_x + C_{12}\varepsilon_y + C_{13}\varepsilon_z + C_{14}\gamma_{xy} + C_{15}\gamma_{yz} + C_{16}\gamma_{zx}$$

$$\sigma_y = C_{21}\varepsilon_x + \ldots$$

$$\sigma_z = C_{31}\varepsilon_x + \ldots$$

$$\tau_{xy} = C_{41}\varepsilon_x + \ldots \qquad (2.15)$$

$$\tau_{yz} = C_{51}\varepsilon_x + \ldots$$

$$\tau_{zx} = C_{61}\varepsilon_x + \ldots$$

in which the terms C_{ij} are the components of the *elastic stiffness* matrix (also known as the *material property* matrix). In matrix notation, (2.15) can be written as $\{\boldsymbol{\sigma}\} = [\mathbf{D}]\{\boldsymbol{\varepsilon}\}$.

The matrix $[\mathbf{D}]$ is symmetric and its 36 terms C_{ij} are functions of the elastic constants for the material. However, some of these terms are often zero. For example, an isotropic, homogeneous material has 12 nonzero terms, which are functions of E and ν only. The zero values occur because:

1. Pure direct stress cannot give shear with respect to the same coordinate, i.e. σ_x is not a function of τ_{xy} or τ_{zx}.
2. Pure shear cannot give tension or compression with respect to the same coordinates, i.e. τ_{xy} is not a function of σ_x or σ_y.

3. Pure shear produces stresses only in the plane of the applied stress, i.e. τ_{xy} is not a function of τ_{yz} or τ_{zx}.

For the isotropic, homogeneous case

$$\varepsilon_x = f(\sigma_x, \sigma_y, \sigma_z) \qquad \gamma_{xy} = f(\tau_{xy}) \tag{2.16}$$

and so on.

For ε_x, one can write

$$\varepsilon_x = S_{11}\sigma_x + S_{12}\sigma_y + S_{13}\sigma_z \tag{2.17}$$

where S_{ij}s are known as *compliances*. Equations (2.13) and (2.14) and experimental data give $S_{11} = 1/E$ and $S_{12} = S_{13} = -\nu/E$. It follows that the three-dimensional stress–strain relationships (2.15) can be written in the matrix form:

$$
\begin{Bmatrix} \sigma_x \\ \sigma_y \\ \sigma_z \\ \tau_{xy} \\ \tau_{yz} \\ \tau_{zx} \end{Bmatrix} = \frac{E}{(1+\nu)(1-2\nu)}
\begin{bmatrix}
1-\nu & \nu & \nu & 0 & 0 & 0 \\
\nu & 1-\nu & \nu & 0 & 0 & 0 \\
\nu & \nu & 1-\nu & 0 & 0 & 0 \\
0 & 0 & 0 & \dfrac{1-2\nu}{2} & 0 & 0 \\
0 & 0 & 0 & 0 & \dfrac{1-2\nu}{2} & 0 \\
0 & 0 & 0 & 0 & 0 & \dfrac{1-2\nu}{2}
\end{bmatrix}
\begin{Bmatrix} \varepsilon_x \\ \varepsilon_y \\ \varepsilon_z \\ \gamma_{xy} \\ \gamma_{yz} \\ \gamma_{zx} \end{Bmatrix}
\tag{2.18}
$$

The material property matrix, [**D**] in (2.18) is for an isotropic material. Equations (2.18) are often presented in the more convenient strain–stress form of

$$\varepsilon_x = \frac{1}{E}[\sigma_x - \nu(\sigma_y + \sigma_z)] \qquad \gamma_{xy} = \tau_{xy}\frac{2(1+\nu)}{E}$$

$$\varepsilon_y = \frac{1}{E}[\sigma_y - \nu(\sigma_x + \sigma_z)] \qquad \gamma_{yz} = \tau_{yz}\frac{2(1+\nu)}{E}$$

$$\varepsilon_z = \frac{1}{E}[\sigma_z - \nu(\sigma_x + \sigma_y)] \qquad \gamma_{zx} = \tau_{zx}\frac{2(1+\nu)}{E}$$

Note that $E/2(1 + \nu)$ is G, the shear modulus of elasticity. Remember, however, that the above relationships (2.18) are valid only when the material is isotropic and the stress is in the linear elastic range.

Formulae in terms of material elastic constants for the C_{ij}s for anisotropic linear elastic materials (e.g. fibre-reinforced composites) and for nonlinear stress–strain behaviour (e.g. plasticity of steels) are also available. Should the finite element analysis require such material stress–strain relationships the user should consult specialist texts and software vendors' manuals.

The values for the elastic constants determine how much deformation occurs. In other words, a structure under a given set of boundary conditions deforms twice as much if the Young's modulus is halved. In certain mechanical designs it is necessary to limit the deformation of the structure, as a deformation greater than some known value is seen as a failure. For example, an automobile body shell must be sufficiently rigid so that the stresses transmitted from it to the windscreen glass are not large enough for the glass to crack; or for a vehicle without a roof, say an open-top sports car, the body must not deflect too much so that the handling of the vehicle is affected.

Although the deformation of a structure changes with choice of material, the stresses at any point remains unaltered as long as deflections are small. This is because stress is a measure of the intensity of force and this is independent of the material. However, all materials have limiting stresses in tension, compression and shear which cause the structure to fail mechanically. These failures can be due to the onset of yielding, gross cross-sectional plasticity (or ductile collapse), fatigue mechanisms and brittle fracture, but they can be accounted for in a design if the strengths of the material and their interactions are known. Table 2.1 presents typical strength data for some structural materials under normal laboratory conditions.

To illustrate how these strengths of materials are determined let us return to the stress–strain curves in Fig. 2.6. The stress–strain behaviour of structural materials differs considerably. As a result the strength of a solid requires careful definition. First, consider a ductile metal (i.e. carbon steels) with a distinct deviation in the stress–strain curve (Fig. 2.6a). The point where this occurs is the *upper yield point*. The stress then drops to a value called the *lower yield point* before increasing again as the strain continues. The corresponding yield strain is usually small, of the order of 0.001 (0.1 per cent). Other metals (e.g aluminium alloys) do not have a clearly defined yield point and their stress–strain curves are as in Fig. 2.6(b). As the *yield stress*, σ_Y (sometimes referred to as *yield strength*), could therefore have a number of definitions it is usually taken to be *0.2 per cent proof stress*. The 0.2 per cent proof stress (Fig. 2.6a) is defined as that stress which results in a 0.2 per cent offset; the stress given by drawing a line parallel to the linear part of the curve through the 0.002 strain value. Also shown in Fig. 2.6(a) is the ultimate tensile strength, σ_U, the ultimate or maximum strength attained during a test, and also listed in Table 2.1.

Deformation at stresses above yield is plastic deformation, and for ductile metals this may be idealized either as perfectly plastic (Fig. 2.6e) or strain hardening (Fig. 2.6f) depending on the behaviour of the material. If the material is stressed beyond the yield point and the stress is removed there will be permanent deformation. Fig. 2.6(b) shows that when the load is reapplied it is found that the yield stress increases. For mechanical design, a knowledge of the strength is of utmost importance. Although σ_U seems to be the most reported strength for ductile materials, the yield or proof stress value is of greater importance as this tells designers when the material is no longer elastic and so they can determine the dimensions of the structure necessary to transmit the applied loads without causing failure. The ultimate strength provides a factor of safety against overload and mechanical failure.

In the mechanical design of structures the design must be fit for its purpose. This means that failure due to the expected loading is to be avoided over the design life of the structure. If the material of the structure is a ductile metal there should not, usually, be any gross cross-sectional yielding. When all, or part, of a structure is transmitting forces predominantly by compressive stresses there is the likelihood that failure of the material is preceded by an instability mode of structural failure, and there are procedures to design against such behaviour. If instability does not occur before material failure then a material failure criterion may be used. There is at present no theoretical way of determining the relationship between the stresses and yielding for a three-dimensional state of stress with yielding in the uniaxial tension test. However, using experimental evidence and mathematical insight two yield criteria have now become generally accepted for predicting the onset of yielding in ductile metals (Dieter, 1984). They are the von Mises' or distortional energy and the maximum shear stress or Tresca's criteria. Both criteria depend on the principal stresses (see Sec. 2.2.2) and yield strength. The former is often a *default* output parameter of commercial finite element software. For this reason it is important to remember that the Cartesian stresses, shown in Fig. 2.2, determined in a finite element analysis can be used to determine the principal stresses and their planes.

The above discussion on material strength applies to metals. For plastics, strength is identified as the stress σ_Y at which the stress–strain curve becomes noticeably nonlinear. Typically, the strain is 0.001 (0.1 per cent). Other engineering materials can be classed as brittle in which little, if any, plastic deformation takes place. If a bar of brittle material such as a unidirectional fibre-reinforced polymer composite is tested, the stress–strain curve shown in Fig. 2.6(c) is obtained, where the curve is practically linear up to the point of ultimate failure when the material yields. In these cases a typical strain at failure can exceed 0.015 and for laminated plates used in the aerospace industry load-carrying capacity is limited by a maximum strain of 0.002.

For metals the compression stress–strain curve up to the yield point is very similar to the tension curve. This is not the case with other structural materials. For ceramics, glasses and concretes the strength depends strongly on the type of loading. In tension the fracture strength is one-tenth of the crushing strength in compression. Table 2.1 quotes some typical values. A full understanding of the strength of a material generally requires the determination of its stress–strain curves in tension, compression and shear. These curves provide several measures of strength according to the relevant mode of failure. Moreover, the material properties of structural materials have a different degree of variability which further complicates the definition of strength. It is the authors' opinion that specialist advice, usually from the material manufacturer or supplier, be obtained on material strength and that values, like Table 2.1, published in texts on selection of materials be used principally for guidance only. It is worth emphasizing that it does not matter how accurate the finite element representation of the structure and its boundary conditions is, the accuracy of the results will only be as good as the material property data available.

2.2.7 Boundary Conditions

The differential equations of equilibrium (2.5) must be satisfied at all interior points in a deformable structure under a three-dimensional force system. The stress components vary over the volume of the structure, and when the boundary (i.e. the surface of the structure) is considered, then these stress components must be in equilibrium with the external forces there. These external forces are always distributed over some area of the boundary such that there is a stress distribution at the surface and these stresses are known as surface tractions. Hence, the external forces may be regarded as a continuation of the internal stress distribution.

If the rectilinear components of the surface forces per unit area are denoted by $\bar{X}$, $\bar{Y}$, $\bar{Z}$, it can readily be shown (Timoshenko and Goodier, 1988) that the boundary conditions for a three-dimensional body are

$$
\begin{aligned}
\bar{X} &= \sigma_x l + \tau_{xy} m + \tau_{zx} n \\
\bar{Y} &= \tau_{xy} l + \sigma_y m + \tau_{yz} n \\
\bar{Z} &= \tau_{zx} l + \tau_{yz} m + \sigma_z n
\end{aligned}
\tag{2.19}
$$

where l, m and n are the direction cosines of the angles that a normal to the surface of the body makes with the x-, y- and z-axes, respectively.

For example, if the boundary of the structure has a plane surface, with an x-directed normal, then $\bar{X} = \sigma_x$, $\bar{Y} = \tau_{xy}$ and $\bar{Z} = \tau_{zx}$. Under this condition the applied surface traction components $\bar{X}$, $\bar{Y}$ and $\bar{Z}$ are balanced by internal stresses σ_x, τ_{xy} and τ_{xz} respectively.

It is usual in the classical methods of stress analysis to specify the loading boundary conditions in terms of surface forces (tractions or pressures), but it is interesting to note that the boundary conditions of a structure may also be given in terms of the displacement components. This is standard practice in a finite element analysis when the element formulation is based on assumed displacement fields (see Chapter 3). When displacement boundary conditions are given, the equilibrium equations (2.5) express the situation in terms of strains, through use of the stress–strain relationships (2.15) and in terms of displacements (2.9).

2.3 SOME SIMPLE SITUATIONS

When introducing the equations of equilibrium (2.5), compatibility (2.12), stress–strain for an isotropic material (2.15) and boundary conditions (2.19), it was stated that the exact solution to a problem must satisfy each of these sets of equations. One of the classical methods for solving the differential equations is to use ordinary differential calculus. Exact solutions for three-dimensional problems are very difficult to obtain because of the large number of unknowns (six stresses, six strains and three displacements) and, often, the complexity of the geometry of the structure. One way of making it easier to obtain a solution is to reduce the number of dimensions of the problem, and hence the unknowns. By doing this solutions for a large number of practical, if simple, structural problems can be obtained. Such solutions have provided the well-known design formulae for structural members such as bars, beams and thin plates.

While the differential calculus method is suitable for problems that are statically determinate, it is often not suitable if the structure is statically indeterminate. In this case a classical solution may be obtained by using an energy method.

2.3.1 One-dimensional Situations and Saint-Venant's Principle

The simplest structure to illustrate the classical solution using differential calculus is a weightless bar, where the body forces are zero, subjected to pure axial loading. This situation models, for small deformation, the gauge length in a tensile (or compressive, providing there is no buckling instability) coupon test discussed in Sec. 2.2.6.

Let the bar be positioned such that its longitudinal centroidal axis lies along the x-axis of a Cartesian system whose origin is at the left-hand end. It has constant circular cross-sectional area A, diameter D, length L, and external load F acting in the x-direction along the centroidal axis. Note that the cross-sectional shape need not be circular but must be constant along the length of the bar.

Equilibrium equations (2.5) are satisfied by taking

$$\sigma_x = \text{constant} = \frac{F}{A} \tag{2.20}$$

and

$$\sigma_y = \sigma_z = \tau_{xy} = \tau_{yz} = \tau_{zx} = 0$$

It is evident that the boundary conditions for the curved surface of the bar at

$$z^2 + y^2 = D^2/4 \qquad x = 0 \text{ to } L \tag{2.21}$$

where there are no external forces, are satisfied, as the only load acts at the ends of the bar. The boundary conditions on the two ends whose surfaces lie in the yz-plane is simply $\sigma_x = \bar{X} = F/A$, from (2.19). Hence, there is a uniform distribution of stress over the cross-section of the bar if the tensile stresses are uniformly distributed over the ends. If this condition is invalidated by the load being applied in some different way, while maintaining the same overall equilibrium (Fig. 2.7), the solution to the problem is no longer exact. As this is the situation in practically every structural problem it would at first appear that the usefulness of the theory is strictly limited. To overcome this difficulty Saint-Venant's principle is invoked. It states:

> While statically equivalent systems of forces acting on a body produce substantially different local effects the stresses at sections distant from the surface of the loading are essentially the same.

Figure 2.7 shows two bars with the same resultant axial load F. Each bar has the same average stress (F/A) at section BB, a distance usually taken to be greater than the minimum dimension of the surface to which the load is applied. However, at section AA closer to the ends, the stress distribution is locally altered owing to the precise nature of the loading. Therefore, the theory may be applied

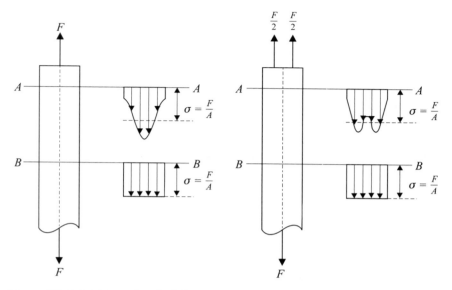

Figure 2.7 Saint-Venant's principle.

to those parts of a structure away from loading surfaces and restraints. In real structures the determination of local stresses which will include higher values than average is best accomplished using a two- or three-dimensional finite element analysis as will be seen in Chapters 7, 8 and 9.

Compatibility equations (2.12) are identically satisfied as the stress–strain equations give

$$\varepsilon_x = \text{constant} = \frac{\sigma_x}{E} = \frac{F}{AE} \qquad (2.22)$$

with the other five strains zero. From (2.9) it is found that, for the one-dimensional situation, $\varepsilon_x = \mathrm{d}u/\mathrm{d}x$. On substitution for the axial strain and integrating along the length of the bar, the familiar expression for the change of length is obtained

$$\Delta L = \frac{FL}{AE} \qquad (2.23)$$

The exact solution for the axially loaded bar uses a modelling approach which lumps the properties of the structure along the centroidal axis, thus effectively reducing the three-dimensional problem to a one-dimensional problem. The same modelling approach is used to provide solutions for the pure torsion of cylinders and tubes, i.e.

$$\frac{T}{I_\mathrm{p}} = \frac{\tau}{r} = \frac{G\theta}{L} \qquad (2.24)$$

where T is the torque, I_p is the polar moment of inertia, τ is the shear stress, r is the radial distance from the axis of twist, G is the shear modulus, θ is the angle of twist per unit length and L is the length of shaft, and for pure bending of beams, i.e.

$$\frac{M}{I_a} = \frac{\sigma}{\pm y} = \frac{E}{R} \tag{2.25}$$

where M is the bending moment, I_a is the second moment of area, σ is the direct stress in the x-direction along the beam, y is the distance from the neutral axis, E is Young's modulus and R is the radius of curvature. The beam solution is not an exact one as it fails to model the secondary deformation that occurs in the plane perpendicular to the beam's length. Equation (2.25) is used to derive the standard bending deflection of beams. If the loading is other than pure bending, for example uniformly distributed loading, then there is addition deformation of the beam owing to the presence of shear.

Each of the three one-dimensional members introduced here is used in the direct matrix stiffness method and the finite element method to analyse structural frames. The latter method did indeed develop from the former in the late 1950s to analyse non-frame-type continuum structures, as discussed in Sec. 1.5.3.

2.3.2 Two-dimensional Situations

In many situations it is possible to consider structures which behave two-dimensionally over much of their volume. For example, consider a structure with a constant cross-section which lies in the xy-plane. With boundary conditions and body forces independent of z the problem is two-dimensional. Now, there are eight unknown quantities (i.e. three stresses, three strains and two displacements), each of which is a function of the x and y spatial coordinates.

With no body forces the equilibrium equations (2.5) become

$$\frac{\partial \sigma_x}{\partial x} + \frac{\partial \tau_{xy}}{\partial y} = 0 \qquad \frac{\partial \tau_{xy}}{\partial x} + \frac{\partial \sigma_y}{\partial y} = 0 \tag{2.26}$$

and the compatibility equations (2.12) become

$$\frac{\partial^2 \gamma_{xy}}{\partial x \partial y} = \frac{\partial^2 \varepsilon_x}{\partial y^2} + \frac{\partial^2 \varepsilon_y}{\partial x^2} \tag{2.27}$$

Also, the strain–displacement relationships from (2.9) are

$$\varepsilon_x = \frac{\partial u}{\partial x} \qquad \varepsilon_y = \frac{\partial v}{\partial y} \qquad \gamma_{xy} = \frac{\partial u}{\partial y} + \frac{\partial v}{\partial x} \tag{2.28}$$

Two linear elastic stress–strain relationships can be developed from the three-dimensional version given in (2.18) depending on the assumed size of the z-dimension.

In plane stress problems of plates or flat sheet material, where the thickness is small compared with the other dimensions, the surfaces of the sheet are frequently free of both normal stresses, σ_z, and shear stresses, τ_{yz} and τ_{zx}. It is therefore assumed that these stress components remain zero throughout the thickness of the sheet, or at least that they are negligibly small compared with the other applied stresses. The state of stress in the sheet is then two-dimensional and the other nonzero stress components σ_x, σ_y and τ_{xy} may be averaged over the thickness. From (2.18)

$$\varepsilon_z = \frac{-\nu(1+\nu)}{(1-\nu^2)}(\varepsilon_x + \varepsilon_y)$$

and

$$\left\{ \begin{array}{c} \sigma_x \\ \sigma_y \\ \tau_{xy} \end{array} \right\} = \frac{E}{1-\nu^2} \left[\begin{array}{ccc} 1 & \nu & 0 \\ \nu & 1 & 0 \\ 0 & 0 & \frac{1-\nu}{2} \end{array} \right] \left\{ \begin{array}{c} \varepsilon_x \\ \varepsilon_y \\ \gamma_{xy} \end{array} \right\} \tag{2.29}$$

In contrast to plane stress, in which the dimension in z is small compared to that in the x- and y-directions, the plane strain condition is characterized by a large dimension in z, for example with long tubes, dams and mines. In this situation it is assumed that the z-displacement component w is zero at every cross-section. The strain components ε_z, γ_{yz}, and γ_{zx} therefore vanish. Moreover, with ε_z set to zero, the stress σ_z can be expressed in terms of σ_x and σ_y. From (2.18)

$$\sigma_z = \nu(\sigma_x + \sigma_y)$$

and

$$\left\{ \begin{array}{c} \sigma_x \\ \sigma_y \\ \tau_{xy} \end{array} \right\} = \frac{E}{(1+\nu)(1-2\nu)} \left[\begin{array}{ccc} 1-\nu & \nu & 0 \\ \nu & 1-\nu & 0 \\ 0 & 0 & \frac{1-2\nu}{2} \end{array} \right] \left\{ \begin{array}{c} \varepsilon_x \\ \varepsilon_y \\ \gamma_{xy} \end{array} \right\} \tag{2.30}$$

and for plane strain problems a slice of unit thickness is usually considered. The general matrix form (2.29) and (2.30) is $\{\boldsymbol{\sigma}\} = [\mathbf{D}]\{\boldsymbol{\varepsilon}\}$, in which matrix $[\mathbf{D}]$ is 3 by 3.

Faced with eight unknowns and eight equations, it is of great advantage to reduce this number. A classical approach is to seek special functions that automatically satisfy some of the equations. The concept of a stress function is now introduced which has the property of representing all the stresses in such a way as to satisfy the equations of equilibrium (2.26). The Airy stress function $\Phi(x, y)$ (Fenner, 1986; Love, 1944; Timoshenko and Goodier, 1988), is defined as being related to the stresses by

$$\sigma_x = \frac{\partial^2 \Phi}{\partial y^2} \qquad \sigma_y = \frac{\partial^2 \Phi}{\partial x^2} \qquad \tau_{xy} = -\frac{\partial^2 \Phi}{\partial x \partial y} \tag{2.31}$$

Starting with the compatibility equation (2.27) and substituting for strains in terms of stresses using either (2.29) or (2.30), followed by substitution for stresses in terms of the Airy stress function Φ, the governing biharmonic equation is obtained

$$\frac{\partial^4 \Phi}{\partial x^4} + 2\frac{\partial^4 \Phi}{\partial x^2 \partial y^2} + \frac{\partial^4 \Phi}{\partial y^4} = 0 \qquad \text{or} \qquad \nabla^4 \Phi = 0 \tag{2.32}$$

Equilibrium, compatibility and stress–strain relationships are satisfied by (2.31) and (2.32). Providing that a stress function Φ, often in the form of a series function, can be found to satisfy the biharmonic equation and the boundary conditions for the problem, the stresses are given by (2.31). Strains and displacements follow from the relevant equations and boundary conditions.

This classical method is restricted to problems with simple geometry and simple loading. However, the restrictions do not detract from the power of the technique to provide designers with relevant solutions to relatively complex problems. Such solutions are used to benchmark the performance of a two-dimensional finite element analysis. This will be illustrated in Chapter 7 using the so-called Brazilian test (or split tensile test) for brittle materials (John, 1992) which consists of a disc with diametrical opposite compression loading.

It is practical to solve simple three-dimensional problems by the classical method but because of the increased number of unknowns the mathematics becomes very complicated.

2.4   DESCRIBING PROBLEMS IN STRUCTURAL MECHANICS

In the preceding sections the theory of linear elastic structures has been considered. Before developing the equations of the finite element method in Chapter 3 it is worth considering the specification of a structural problem that must be made regardless of the analysis tool to be used. Readers should be aware that finite elements are not always the best analytical tool for a given problem. For example, an approximate but usable solution might well be found by the application of some standard formula. Equally, finite difference methods are ideal for shells of revolution and boundary methods are ideal for problems with boundaries at infinity. It is through careful specification of the problem that the analyst should have sufficient information to be able to decide upon the best analysis tool for the particular problem under consideration.

When thinking about the structural problem, the analyst must determine a specification of the problem. This should be a clear exposition of the reasoning that is driving the analysis, outlining what information is required at the end of the analysis and as much as is known about the physical problem itself. This specification then becomes the main source of information for the analyst.

2.4.1   Producing a Specification

A specification for a structural problem must be sufficiently detailed so that the analyst can obtain from it all the information necessary to define the problem. This information comes from a good understanding of the engineering problem which the analyst must obtain by discussion with the people who require the results of the simulation. In particular the analyst must know the following:

● the requirements of the analysis
● the geometry of the problem

- the possible range of materials to be used
- the physical loads and restraints on the structure.

Requirements of the analysis   Analysing a structure can be an expensive business. If someone wants to commission a computational analysis then considerable expense will probably be involved as access to computer hardware must be provided along with the necessary software and highly trained analysts to produce the solution. Even if computers are not used the analysts must still be paid and, consequently, there must be good reasons for carrying out the analysis. The analysts must therefore explore these reasons first, by talking to the people who need the results of the simulation, such as design engineers.

The reasons for an analysis are many and varied. In Sec. 1.4 examples of where computers might be used to calculate the behaviours of structures were considered, i.e. strength, displacement, effects of heat, optimum material thicknesses, fatigue, dynamic response and crash worthiness. Further, each individual application has its own peculiarities. For example, there might be known limits for stress or displacement determined from experience or to meet legislative requirements, and analysts must know the appropriate form of the results that the simulation should produce; often they must do this from only a vague description of an engineering problem.

Once analysts know the reasoning behind the structural problem it is easier to plan ahead so that any computational model produces the necessary information. One further benefit of this discussion between analysts and their clients is that they get to know each other and their respective problems. Such an understanding can help the analysis process to be brought to a successful conclusion, especially if things do not quite go as planned.

At the end of this initial part of the specification phase analysts should have a list of the data that the computational model must produce. This could include the following:

- the peak stresses in the structure
- the displacement of certain parts of the structure
- the fatigue life of the structure
- the time history of the deformation or damage growth
- the energy absorbed during deformation.

Clearly, assessment of the suitability of results should be made with the clients.

At this stage it might be that the analysts judge that computational methods are not suitable for the physical problem being considered, perhaps because of the unreliability of numerical methods in this case, or because the cost is prohibitive when compared to physical experiment, for example. It is better to make this clear to the clients at this stage so as to prevent a loss of face or embarrassment later. Some pointers as to the suitability of numerical methods will be given in Chapters 7–9.

Specifying the geometry of the structure When looking at any structural problem it is important to be able to describe the geometry of the structure either in detail or as an approximation.

Various sources of geometrical data can be available and these can be used by the analyst to describe the bounding surfaces. For example, this data might come from:

- analytical descriptions of shapes in two dimensions given by such things as points, lines, arcs and splines
- engineering drawings
- databases created by computer-aided design (CAD) systems
- physical three-dimensional models
- measurements taken from existing components.

From such sources most of the bounding surfaces of the geometry may be determined precisely. This information might be used when building the mesh of points inside the domain, but during the specification stage it is sufficient to know roughly where these surfaces are in relation to each other and how they fit together. A simple sketch might help to show this.

Determining the material properties Having developed an understanding of the geometry of the problem the actual material properties for the structure need to be found. These might include the elastic constants, Young's modulus and Poisson's ratio, the material strengths and the effects of anisotropy as outlined in Sec. 2.2.6.

Defining the applied loading and restraints Once the geometry and material properties of the problem are understood the analyst must think about the way in which the structure will be loaded for the particular analysis. In mathematical terms this involves defining appropriate boundary conditions for the problem and, depending on the analysis, initial conditions. The first step here is to determine the external loads themselves in terms of their values as concentrated forces and pressures and the positions of their application on the structure. Then the positions where the structure is supported must be determined. Possibilities include a full restraint support where all displacements are set to zero, or a sliding restraint support where the motion of the boundary is constrained to be in a plane or along a line.

2.4.2 Physical Behaviour of the Structure

From all of the above information it should be possible to sketch out how the structure will behave physically. In this text only linear elastic small displacement problems are considered in the main, and so consideration must be given in design to the structural response having buckling instabilities and nonlinearities leading to large deflection and plasticity. If these are present in regions of interest then appropriate action should be taken as will be discussed in Chapter 10.

However as a guide:

> If nonlinearities are localized and occur away from the regions of interest then the analyst is at liberty to ignore the behaviour and treat the problem as small displacement linear elastic.

Note also that a static analysis will provide results independent of the real behaviour of the structure and its material. In other words, if the loads are high enough the analysis will output displacements and stress values that would trigger, if activated, a nonlinear or instability response in the structure. For example, a long slender solid steel rectangular-shaped column, fixed at one end, free at the other end and subjected to a pure axial compressive load, i.e. there is no moment, is a classical bifurcation buckling problem of Euler. However, a static analysis of this problem results in ever-increasing displacement and stress as the load increases and there is no buckling when the load passes the critical Euler buckling value. Equally, stresses will continue to increase even when they exceed, many fold, the yield (and ultimate) strength of the steel.

Note that when modelling, real imperfections that always exist in a structure and its loading and supports are often ignored completely. Some account of this simplification needs to be made if buckling instabilities are a likely mode of failure.

2.5 REFERENCES AND FURTHER READING

In his classic book, Gordon (1976) analyses structural problems in some detail and in a readable style. Standard and more mathematical texts on the derivation of the governing equations and classical solutions are Love (1944), Timoshenko and Goodier (1988), Gere and Timoshenko (1991) and Fenner (1986). However, a useful summary of the formulae, facts, and principles pertaining to the classical methods is given in the compendium commonly known as *Roark's Formulas for Stress and Strain* Young (1989).

Material properties are discussed by John (1992) and Dieter (1986). However, anisotropy is covered by Calladine and Christopher (1985), Dieter (1986), Halpin (1992) and software vendors' manuals. Finally, the selection of materials is discussed by Ashby (1992) and Charles and Crane (1989).

3

FINITE ELEMENT SOLUTIONS OF THE EQUATIONS

Chapter 1 discussed structural design problems together with the historical development of the finite element method, while Chapter 2 discussed the governing equations for linear elastic structures. Consequently, the mathematical techniques for producing numerical solutions on a computer can now be reviewed.

Our starting point is energy methods, as more than 88 per cent of the continuum elements are derived by applying the principle of minimum potential energy over an element. The way the displacement of a structure is approximated is crucial to the finite element formulation, and so some time will be spent discussing these distributions. Direct methods of element equation formulation will also be discussed.

Whatever method is used to form the element equations, integration of functions must be performed through an element. This can be done analytically for elements of simple shape, but for many elements numerical integration must be used and so this is also discussed before looking at special element types. Finally, the ways in which the numerical solution of the equations is produced are discussed.

3.1 DEVELOPING THE FINITE ELEMENT METHODOLOGY

3.1.1 An Expression for Potential Energy

When an elastic structure is loaded by some external forces or moments its potential energy is increased. This potential energy is made up of the internal strain energy due to deformation and the potential of the loads that act within the structure (e.g. body forces) or on its surface. The strain energy is generated as atoms are moved from their equilibrium positions as a result of deformation of the structure.

Some mathematical way of describing this process must now be found. In Chapter 2 the stress–strain relationship was defined, see (2.15) and (2.18), of which

$$\{\boldsymbol{\sigma}\} = [\mathbf{D}](\{\boldsymbol{\varepsilon}\} - \{\boldsymbol{\varepsilon}_O\}) + \{\boldsymbol{\sigma}_O\} \tag{3.1}$$

is a modified form to account for any initial strains ε_O and stresses σ_O.

As strain energy is defined mathematically as one half of the product of stress and strain per unit volume, (3.1) can be used to develop an expression for this energy. By considering a linear elastic structure that carries conservative loads (i.e. the work done by the loads is path independent), and has volume V and surface area S, the general expression for the potential energy, Π_p, can be written as the functional (Washizu, 1982):

$$\Pi_\mathrm{p} = \int_V \left(\frac{1}{2} \{\boldsymbol{\varepsilon}\}^\mathrm{T}[\mathbf{D}]\{\boldsymbol{\varepsilon}\} - \{\boldsymbol{\varepsilon}\}^\mathrm{T}[\mathbf{D}]\{\boldsymbol{\varepsilon}_O\} + \{\boldsymbol{\varepsilon}\}^\mathrm{T}\{\boldsymbol{\sigma}_O\} \right) \mathrm{d}V$$
$$- \int_V \{\mathbf{u}\}^\mathrm{T}\{\mathbf{X}\}\mathrm{d}V - \int_S \{\mathbf{u}\}^\mathrm{T}\{\bar{\mathbf{X}}\}\mathrm{d}S - \sum_{p=1}^{P}\{\Delta_p\}^\mathrm{T}\{\mathbf{R}_p\} \tag{3.2}$$

where, for a Cartesian coordinate system:

$\{\mathbf{u}\}$ is $[u \ v \ w]^\mathrm{T}$, the displacement components throughout the structure (Sec. 2.2.4).

$\{\boldsymbol{\varepsilon}\}^\mathrm{T}$ is $[\varepsilon_x \ \varepsilon_y \ \varepsilon_z \ \gamma_{xy} \ \gamma_{yz} \ \gamma_{zx}]$, the strain components, given in terms of displacements by (2.10).

$[\mathbf{D}]$ is the material property matrix, for example, $[\mathbf{D}]$ in (2.18) is for an isotropic material.

$\{\boldsymbol{\varepsilon}_O\}^\mathrm{T}$ is $[\varepsilon_x^O \ \varepsilon_y^O \ \varepsilon_z^O \ \gamma_{xy}^O \ \gamma_{yz}^O \ \gamma_{zx}^O]$, the initial strain components due to, for example, free expansion, shrinkage or moisture swelling.

$\{\boldsymbol{\sigma}_O\}^\mathrm{T}$ is $[\sigma_x^O \ \sigma_y^O \ \sigma_z^O \ \tau_{xy}^O \ \tau_{yz}^O \ \tau_{zx}^O]$, the initial stress components due to, for example, pre-stressing or residual stresses.

$\{\mathbf{X}\}^\mathrm{T}$ is $[X \ Y \ Z]$, the body forces per unit volume (Eqs 2.5).

$\{\bar{\mathbf{X}}\}^\mathrm{T}$ is $[\bar{X} \ \bar{Y} \ \bar{Z}]$, the surface tractions (Sec. 2.2.7 and Eqs 2.19).

$\{\Delta_p\}$ is the vector of P translational and rotational displacements corresponding to their P external concentrated forces and moments in vector $\{\mathbf{R}_p\}$.

Now it can be seen that in (3.2) there are six terms. The first comes from (3.1) and is the main term as it accounts for the potential energy due to the current levels of stress and strain. The second and third terms also come from (3.1) and these account for the energy changes due to any initial stresses and strains. The fourth term represents the energy due to the presence of body forces, the fifth term that due to surface tractions and the final term that due to any external concentrated forces and moments.

Note that the integrals in (3.2) that contain $\{\mathbf{X}\}$ and $\{\bar{\mathbf{X}}\}$ are the work done (hence potential lost) by body forces and surface tractions as the structure deforms. Integration is performed only over the portions of volume V and surface area S where body forces and surface tractions act. They (i.e. the integrals in (3.2) that contain vectors $\{\mathbf{X}\}$ and $\{\bar{\mathbf{X}}\}$, and the potential energy calculated after solving these integrals), can be regarded as potential changes of $-Xu - Yv - Zw$ per unit volume and $-\bar{X}u - \bar{Y}v - \bar{Z}w$ per unit surface area, respectively. The final term represents the potential lost by concentrated forces and/or moments moving as the structure deformed and it can be written as $-\Delta_1 R_1 - \Delta_2 R_2 - \Delta_3 R_3 \ldots - \Delta_P R_P$.

3.1.2 The Rayleigh–Ritz Method

Equation (3.2) relates the potential energy of a deformed structure to the strains within the structure and to the initial stresses and strains, the applied external forces, the body forces, the surface tractions and the material properties. Clearly, everything except the strains and stresses will be known from the definition of the structural problem.

When any structure is deformed, the continuum that makes up the structure moves from the equilibrium position to some other position. To describe this deformation in full the displacement at all positions within the continuum must be known, and it is said that the problem has an infinite number of degrees of freedom, i.e. an infinite number of points as discussed in Sec. 2.2. To make the problem amenable to solution, the number of degrees of freedom must be reduced to a manageable number. This can be done using the Rayleigh–Ritz method.

This method involves the approximation of the displacement components u, v and w for the whole structure by distributions which have the form

$$u = \sum_{i=1}^{l} a_i f_i(x, y, x) \qquad v = \sum_{i=l+1}^{m} a_i f_i(x, y, z) \qquad w = \sum_{i=m+1}^{n} a_i f_i(x, y, z) \qquad (3.3)$$

in which the coefficients a_i, whose values are yet to be determined, are known as *generalized coordinates* . The functions f_i are known as *admissible functions* and are usually polynomials. For a function to be admissible it must satisfy compatibility conditions, Eqs (2.12) and displacement boundary conditions. It is not required that any of the functions f_i satisfy exactly stress (loading) boundary conditions, but if they do so then accuracy is improved. Here, the analyst must determine both the number of terms and the form of the functions in order to achieve the desired solution accuracy. Hence the unknowns are simply the generalized coordinates a_i.

To determine the n unknown generalized coordinates the following procedure is applied. The assumed displacement distributions are substituted into the strain–displacement expressions (2.9) to find strains $\{\varepsilon\}$, then the functional (3.2) is used to evaluate Π_p. Thus the potential energy of the structure becomes dependent on the coordinates a_i alone. Now the principle of minimum potential energy is applied, as discussed in Sec. 1.5.3. To find the minimum energy the differential of Π_p with respect to the all generalized coordinates must be set to zero to find the stationary values. Hence the deformed equilibrium configuration of the structure in terms of n algebraic equations is derived from the minimization as

$$\left\{ \frac{\partial \Pi_p}{\partial a_i} \right\} = \{0\} \qquad \text{for} \qquad i = 1, 2, \ldots, n \qquad (3.4)$$

Equations (3.4) are found to be stiffness equations and can be written in the matrix form $[\mathbf{K}]\{\mathbf{a}\} = \{\mathbf{F}\}$. This form is similar to Eq. (1.2), except that the unknown degrees of freedom are generalized coordinates and not nodal displacements. Numerical values for the coordinates a_i are obtained from solving (3.4). When substituted into the expressions in (3.3) they provide the approximate displacement distributions in the structure. Differentiation of the displacements u, v and w yields strains which, when substituted into the stress–strain relationships, will give stresses.

3.1.3 Using Local Displacement Distributions

Recall from the historical development in Sec. 1.5 that, by the 1940s, engineers and mathematicians had realized that it was not possible to find admissible functions, f_i, capable of describing the exact deformation when a structure had general geometry, boundary conditions and material properties. Such a realization was the catalyst for developing different numerical methods that treated a structure as a sum of smaller subregions (or elements).

As integration can be thought of as a summation process, it follows that by discretizing the structure into a finite number of elements and summing each of their potential energy contributions the method will approximate the total potential energy, as given by the single body functional (3.2). When implementing this method, the finite element method, the potential energy functional (3.2) must be rewritten as a summation over all the elements that make up the structure. The last term in (3.2) is not amenable to this and so it is treated as follows.

The displacement vector $\{\Delta_p\}$ in the last term is replaced by the vector of all displacements at all nodes (the degrees of freedom) in the model $\{\delta\}$. Note that nodes should be placed where the external concentrated forces and/or moments have been applied so that the displacements in $\{\Delta_p\}$ can be readily expressed in terms of nodal displacements. The vector of applied forces and moments $\{\mathbf{R}_p\}$, is replaced by the global vector of nodal applied forces, $\{\mathbf{R}\}$, which has terms associated with each of the displacements in $\{\delta\}$. Note that the forces and/or moments in $\{\mathbf{R}_p\}$ may have to be transformed before they are inserted into $\{\mathbf{R}\}$, such that the energy terms $-\sum_{p=1}^{P}\{\Delta_p\}^{\mathrm{T}}\{\mathbf{R}_p\}$ and $-\sum_{\delta=1}^{n}\{\delta\}^{\mathrm{T}}\{\mathbf{R}\}$ are equivalent.

Many terms in both $\{\delta\}$ and $\{\mathbf{R}\}$ will be zero, as nonzero terms are generated only where there are external applied loads.

Having transformed the last term in (3.2), the element form of the remaining terms can be considered. For the single body problem in Sec. 3.1.2 it was seen that the deformation could be approximated, although not necessarily accurately, by assuming the Rayleigh–Ritz displacement distributions of expressions (3.3). Now a similar approach is needed for a single element to provide admissible simple displacement distributions that can be used to generate the element characteristics, the element stiffness matrix $[\mathbf{k}]$ and the element force vector $\{\mathbf{F}^e\}$. For now it is sufficient to assume that the appropriate displacement distributions for an element can be interpolated from the element nodal displacements $\{\boldsymbol{\delta}^e\}$. This can be defined in matrix form as

$$\{\mathbf{u}\} = [\mathbf{N}]\{\boldsymbol{\delta}^e\} \tag{3.5}$$

where $[\mathbf{N}]$ is the shape function matrix, standard forms of which for various elements will be presented later.

Now, the derivation of the element representation of (3.2) can be completed as follows. In Chapter 2 the relationship between strains and displacements is given by (2.9) and in matrix form by (2.10). This can be rewritten as

$$\{\boldsymbol{\varepsilon}\} = [\boldsymbol{\partial}]\{\mathbf{u}\} \tag{3.6}$$

in which $[\boldsymbol{\partial}]$ is the differential operator matrix which, for a three-dimensional problem, is given by (2.10). Definition of the element strains in terms of the element nodal displacements is, therefore, given by

$$\{\boldsymbol{\varepsilon}\} = [\mathbf{B}]\{\boldsymbol{\delta}^e\} \qquad \text{where} \qquad [\mathbf{B}] = [\boldsymbol{\partial}][\mathbf{N}] \tag{3.7}$$

when (3.5) is substituted into (3.6).

Substitution of the element expressions (3.5) and (3.7) for $\{\mathbf{u}\}$ and $\{\boldsymbol{\varepsilon}\}$ into the functional (3.2), and summing the contributions from M elements, gives the potential energy of the structure as

$$\Pi_{\mathrm{p}} = \frac{1}{2}\sum_{e=1}^{M}\{\boldsymbol{\delta}^e\}^{\mathrm{T}}[\mathbf{k}]\{\boldsymbol{\delta}^e\} - \sum_{e=1}^{M}\{\boldsymbol{\delta}^e\}^{\mathrm{T}}\{\mathbf{F_q}^e\} - \{\boldsymbol{\delta}\}^{\mathrm{T}}\{\mathbf{R}\} \tag{3.8}$$

Here, the first term comes from the first term in (3.2), the second term is a composite of the second, third, fourth and fifth terms in (3.2) and the final term is in the modified form discussed previously in this section. Hence it can be seen that from (3.8) the element stiffness matrix is defined by

$$[\mathbf{k}] = \int_{V^e} [\mathbf{B}]^{\mathrm{T}}[\mathbf{D}][\mathbf{B}]\mathrm{d}V \tag{3.9}$$

and that the *consistent* (or *kinematically equivalent*) element force vector is defined as

$$\{\mathbf{F_q^e}\} = \int_{V^e}[\mathbf{B}]^{\mathrm{T}}[\mathbf{D}]\{\boldsymbol{\varepsilon}_O\}\mathrm{d}V + \int_{V^e}[\mathbf{B}]^{\mathrm{T}}\{\boldsymbol{\sigma}_O\}\mathrm{d}V + \int_{V^e}[\mathbf{N}]^{\mathrm{T}}\{\mathbf{X}\}\mathrm{d}V + \int_{S^e}[\mathbf{N}]^{\mathrm{T}}\{\bar{\mathbf{X}}\}\mathrm{d}S \tag{3.10}$$

where V^e and S^e are the volume and surface area of an element respectively.

Equation (3.10) shows how certain distributed loads can be transformed into equivalent nodal forces associated with the nodal displacements used to describe the deformation of the structure. The word 'consistent' is used to indicate that the transformation is consistent with the shape functions N_i of (3.5) used to generate [**k**], and that under certain conditions (see Sec. 3.7.7) the work done by the actual loading is equal to that in the finite element analysis. To complete an element force vector in (1.1), the external concentrated forces and/or moments from $\{\mathbf{R}\}$ may be added to the consistent force vector to yield

$$\{\mathbf{F}^e\} = \{\mathbf{R}^e\} + \{\mathbf{F}_q^e\} \tag{3.11}$$

Usually this step is done when the characteristics of each element are assembled into the governing simultaneous equations. Equation (3.11) has been given here to show that an element contribution to $\{\mathbf{F}\}$ has two different components and is not due to (3.10) alone.

Of the two element characteristics the expression for [**k**] is the more important. Although the expression (3.10) for $\{\mathbf{F}_q^e\}$ looks formidable in typical static problems, many of the terms $\{\boldsymbol{\varepsilon}^O\}$, $\{\boldsymbol{\sigma}^O\}$, $\{\mathbf{X}\}$, $\{\bar{\mathbf{X}}\}$, contributing to the consistent force vector are zero. Consequently, body forces, initial strains and initial stresses are generally ignored in the rest of the text to simplify the presentation. To find out how all these energy terms alter the formulation of $\{\mathbf{F}_q^e\}$ interested readers are referred to the general texts in Sec. 3.10.

In making this simplification expression (3.10) reduces to a single term for the surface tractions:

$$\{\mathbf{F}_q^e\} = \int_{S^e} [\mathbf{N}]^T \{\bar{\mathbf{X}}\} dS \tag{3.12}$$

To complete the element derivation the algebraic equations that are solved to yield the displacements must be obtained. Every displacement in the element vector $\{\boldsymbol{\delta}^e\}$ also exists in the global (or structure) vector $\{\boldsymbol{\delta}\}$ of displacements. As a consequence of the assembly process which must include each of the M element contributions, the last two terms in (3.8) can be combined to give

$$\Pi_p = \frac{1}{2} \{\boldsymbol{\delta}\}^T [\mathbf{K}] \{\boldsymbol{\delta}\} - \{\boldsymbol{\delta}\}^T \{\mathbf{F}\} \tag{3.13}$$

where

$$[\mathbf{K}] = \sum_{e=1}^{M} [\mathbf{k}] \quad \text{and} \quad \{\mathbf{F}\} = \sum_{e=1}^{M} \left(\{\mathbf{F}_q^e\} + \{\mathbf{R}^e\} \right) \tag{3.14}$$

Now, minimization of Π_p to find its stationary state with respect to a small variation in each of the nodal displacements δ in $\{\boldsymbol{\delta}\}$ gives

$$\left\{ \frac{\partial \Pi_p}{\partial \boldsymbol{\delta}} \right\} = \{\mathbf{0}\} \tag{3.15}$$

the solution to which is

$$[\mathbf{K}]\{\boldsymbol{\delta}\} = \{\mathbf{F}\}$$

i.e. (1.2) again. Solving the n simultaneous equations in (1.2) yields the n degrees of freedom in $\{\boldsymbol{\delta}\}$. After these primary unknown displacements have been determined, strain and stresses in an element may be calculated from (3.7) and

$$\{\boldsymbol{\sigma}\} = [\mathbf{D}][\mathbf{B}]\{\boldsymbol{\delta}^{\mathbf{e}}\} \qquad (3.16)$$

respectively.

It is important to appreciate that the process used in the finite element method, as given by (1.2), is universal and is therefore independent of the nature of any static element type used to model a structural problem.

3.2 SATISFYING EQUILIBRIUM AND COMPATIBILITY

Chapter 2 discussed what must be achieved for an exact solution of a static problem. It was shown that in an elasticity solution every small subregion enclosing a volume of material is in static equilibrium (Sec. 2.2.3) and compatibility is satisfied throughout (Sec. 2.2.5). In Sec. 3.1 a finite element methodology was developed which discretized the problem by using displacement distributions to interpolate nodal unknowns within an element. Therefore it is expected that an approximate finite element solution will not meet the necessary requirements in every sense. The degree to which equilibrium and compatibility may be satisfied at nodes, across interelement boundaries and within each element is as follows:

1. *Equilibrium of nodal forces and moments is satisfied* Structural equations $[\mathbf{K}]\{\boldsymbol{\delta}\} = \{\mathbf{F}\}$ are equilibrium equations at nodes. Therefore, the solution for nodal displacements is such that nodal forces have zero resultant at every node.
2. *Compatibility prevails at nodes* Elements are compatible at nodes to the extent of the nodal degrees of freedom they share. This general statement allows for hinge or roller connections and is not restricted to permanent connection as is the physical situation for nodes representing a point in the continuum.
3. *Equilibrium is usually not satisfied across interelement boundaries* When examining element stress results, the stresses in adjacent elements do not necessarily have the same values along the common boundary or at common nodes. For a properly designed mesh these discrepancies are small and can become insignificant when the mesh is refined in the limit.
4. *Compatibility may or may not be satisfied across interelement boundaries* Incompatibility between elements, when it occurs, should tend to zero as more and more elements are used to model a structure. Indeed, this *must* be true if such an element is to solve real problems.

5. *Equilibrium is usually not satisfied within elements.* In general, to satisfy the differential equations of equilibrium (2.5), at every point in an element demands a relationship between the nodal displacements $\{\delta^e\}$ that usually does not exist from the solution of (1.2). The exact solution does exist for certain element types as the size of the element becomes smaller and smaller.

6. *Compatibility is satisfied within elements.* This is achieved if an element displacement distribution, as defined by (3.5), is continuous and single-valued.

The first four of these conditions result from the process of discretizing the continuum—the structure—into elements, whereas the last two conditions relate to the elements themselves. It is important that any inherent discretization error is not increased significantly, but conditions 3 and 4 infer that the accuracy of solution is a function of the mesh construction. However, conditions 5 and 6 lead to a number of requirements that should be satisfied when choosing the polynomial expressions that define an element's displacement distributions. This will be discussed further in Sec. 3.5.2.

3.3 USE OF ELEMENTS WITH DIFFERENT SPATIAL DIMENSIONS

When modelling any structural problem, the geometry must be split into a variety of elements. To do this, elements are used which essentially have one of the five basic forms shown in Table 3.1. These range from a single lumped mass to a three-dimensional volume and are used to model the situations listed in the table. Programmed within the largest commercial finite element codes, which can model various elasticity problems (both linear and nonlinear) and a variety of other field problems, such as thermal conduction, might be more than 100 different elements. A typical range of such elements are shown in Table 3.2. The

Table 3.1 The basic element geometries

Dimensionality	Type	Geometry
Point	Mass	
Line	Spring, beam, bar, spar, gap, torsion	
Area	2D continuum, axisymmetric continuum, plate or flat shell	
Curved area	Generalized shell	
Volume	3D continuum	

Table 3.2   Typical range of elements available in commercial software packages

Element type	Degrees of freedom	Representation
Mass	–	
2D bar	u, v	
2D beam	v, θ_z	
2D continuum plane stress plane strain axisymmetric	u, v	
2D interface	u, v	
Axisymmetric shell	u, v, θ_z	
3D bar	u, v, w	
3D beam	v, v, w $\theta_x, \theta_y, \theta_z$	
3D solid	u, v, w	
3D shell	u, v, w $\theta_x, \theta_y, \theta_z$	
3D interface	u, v, w	

number and nature of the nodal degrees of freedom, both displacement and rotation, for each element are given in the table. Note that the rotation (or slope) degrees of freedom for plate bending and shell elements are obtained by differentiation of a displacement component.

In the physical world all structural forms must be three-dimensional, so why are one- and two-dimensional elements found in finite element codes? To answer this some aspects of modelling must be considered.

The simplest two-dimensional element has a triangular shape and three corner nodes. Its three-dimensional equivalent is a tetrahedron with four corner nodes. Both element types are shown in Table 3.2. If it is assumed that a situation can be modelled as a plane strain problem either element type can be used, but the computational work involved is vastly different. To appreciate the relative magnitude of the computation when solving the simultaneous equations (1.2) for a problem meshed with three-dimensional elements compared to two-dimensional elements, let us consider a simple example. Here, it is assumed that the accuracy of both element types are comparable.

If an adequate analysis of a problem with two-dimensional elements requires a mesh with 20 nodes in each direction, i.e. 400 nodes in total, the total number of simultaneous equations is 800 as there are 2 degrees of freedom per node. Further, the half-bandwidth (see Sec. 3.9) of the matrix $[\mathbf{K}]$ in (1.2) is 20 nodes or 40 variables. For an equivalent three-dimensional representation a cube must be used with 20 nodes in each direction, i.e. 8000 nodes in total. Now the total number of simultaneous equations is 24 000 as each node has 3 degrees of freedom. Furthermore, the bandwidth is now 20 nodes squared (400 nodes) or 1200 variables. Given that with the usual solution techniques discussed in Sec. 3.9 the computational effort is roughly proportional to the number of equations and to the square of the bandwidth, the computational effort required to solve the three-dimensional problem is some 27 000 times greater. There are further disadvantages to using a three-dimensional mesh as there are inherent round-off errors which grow as the number of degrees of freedom increases. Hence, it is not surprising that analysts prefer to model the problem with the element type that, for the specified accuracy needed to design, requires the least amount of preparation time and computational resources. Sometimes the only suitable element type is three-dimensional and then the analyst must decide how many degrees of freedom are needed for an acceptable model and compare this to the number that the code and hardware can handle.

It follows from this that, in commercial packages, there are many elements that are one- or two-dimensional. Such elements represent certain three-dimensional forms where simplifications can be made to reduce the dimension of the form. This modelling approach was shown in Sec. 2.3.1 to yield one-dimensional solutions. For a bar member it is found that (2.23), giving the change in length, satisfies the theory of elasticity exactly, whereas in the case of the beam member, although the solution (2.25) is not exact, the secondary deformation neglected in the theory is of little consequence when analysing a structure of slender beams. Similarly, in Sec. 2.3.2 an in-plane loaded structure is modelled as a two-dimensional body by making the assumption that either the out-of-plane direct strain or stress is zero.

3.4 OVERVIEW OF METHODS FOR CALCULATING ELEMENT EQUATIONS

From Eq. (1.1) for any element it can be seen that for each element the stiffness matrix $[\mathbf{k}]$ and nodal force vector $\{\mathbf{F}^e\}$ will have a different number of terms.

For static analysis the matrix $[\mathbf{k}]$ is formulated using (3.9), with $\{\mathbf{F^e}\}$ being obtained from (3.10) with the addition of the element concentrated forces from vector $\{\mathbf{R}\}$ as shown in (3.11). It can be seen from (3.9) and (3.10) that an element formulation needs definitions of the three basic matrices $[\mathbf{N}]$, $[\mathbf{B}]$ and $[\mathbf{D}]$. As has been seen in Table 3.2, there are numerous classes of element and for each class there are a number of different ways to derive their element characteristics.

There is not scope here to cover the formulation of all types element, and so there will be a concentration on the fundamental features of how element characteristics are derived, placing emphasis on those features that affect accuracy. Sections 3.4–3.8 give sufficient background to enable the reader to refer to, and understand, specialized texts devoted to element formulation.

To derive the matrix $[\mathbf{k}]$ and the vector $\{\mathbf{F^e_q}\}$ a variety of methods may be used. One method is to apply physical reasoning, and this gave the direct stiffness method before computational methods were available. Unfortunately, it is limited to a few one-dimensional element types and so will not be discussed further.

Another method, known as the *direct approach*, considers the integrals in (3.9) and (3.10) directly. To formulate these integrals this method uses polynomial displacement distributions (known as displacement functions) which are dependent on a local element coordinate system or on a global Cartesian coordinate system. This approach was used by the early finite element method pioneers in the 1950s and 1960s when formulating the first two-dimensional elements, and will be discussed in detail in Sec. 3.6.

In two- and three-dimensional problems elements can have complicated geometries with curved boundaries. In these cases it is very difficult, if not impossible, to define appropriate displacement functions over such shapes to satisfy the requirements necessary for accuracy in the finite element method. To overcome this difficulty a method using isoparametric elements was developed by Taig in 1958 (Robinson, 1985). This is the method used in most commercial software as it can routinely deal with elements having curved sides and/or surfaces (see Sec. 3.7).

Attempts have also been made to formulate the element equations using assumed stress distributions and the *principle of complementary energy* (Washizu, 1982). This method overestimates the strain energy and therefore give an *upper bound* solution with underestimated stresses that are safe for design purposes. However, pure equilibrium elements, as they are known, are difficult to create for all but the simplest element types and it is considered that further development in this field is unlikely within commercial software.

A compromise is the formulation of *stress-hybrid elements* that are based neither on the pure displacement method nor on the pure equilibrium method. Here, it has been found feasible to mix the two methods both inside an element and across interelement boundaries. A number of finite element packages now offer a few hybrid elements and they have been found to perform well. However, as it is more likely that packages have displacement elements, this text concentrates on their formulation and accuracy.

3.5 THE ROLE OF DISPLACEMENT DISTRIBUTIONS

3.5.1 Shape Functions

Formulation of the matrix $[\mathbf{k}]$ and the vector $\{\mathbf{F_q^e}\}$ is crucially dependent on the local shape functions in (3.5) that are used in interpolating the nodal displacements throughout the element. In fact, there is not a free choice of terms N_i in a shape function matrix $[\mathbf{N}]$, as for some choices the method fails. In practice the choice of these functions for a specific element type is further restricted, if they are to satisfy all, or most, of the requirements given in Sec. 3.5.2. It is also clear that, whatever method is used to create them, the shape functions must have unit value at node i and zero value at all other nodes on the element. This ensures that the displacements at each node have the correct value.

An element's displacement distribution is nonzero (except at isolated points) in only the small part of the continuum which forms the element, being zero everywhere else in the structure. It is this property that produces so many zeros in the matrix elements of the global stiffness matrix $[\mathbf{K}]$. This matrix is said to be sparse providing assembly has been optimized and this allows relatively efficient solutions of the very large set of simultaneous equations to be found.

Using shape functions leads to an approximation for the displacements over most of the element and it is hoped that this does not differ too much from reality. It can be shown that the strain energy in a structure, found using (3.8), is less than the true strain energy. Moreover, as the number of degrees of freedom is increased, the calculated strain energy approaches the exact value. A solution by the displacement finite element method is therefore referred to as a *lower bound* solution, or a solution in which the structure is found to be *over-stiff*. This over-stiff solution is valid globally but does not apply to every point in the structure. Stresses are overestimated when the numerical method has an over-stiff representation and this is not really what the analyst likes when optimizing a mechanical design.

3.5.2 Requirements to be Met by Displacement Distributions

Certain requirements must be satisfied by any assumed displacement distribution for the finite element method to work, and for many element types this is easily achieved. These requirements are given now since these underpin all element formulations based on the integrals in (3.9) and (3.10).

1. *Number of terms in the polynomial* The number of terms in a polynomial expression selected to represent the unknown displacements must at least equal the number of degrees of freedom associated with the element.
2. *Differentiability* The assumed displacement distribution and its derivatives should be continuous within an element. Since simple polynomial expressions are usually used, no difficulty occurs here and this requirement is inherently satisfied.

3. *Rigid-body modes* The assumed displacement distributions should allow rigid body displacements without invoking strain in an element. To achieve this requirement all polynomial expressions must include a constant term.

4. *Constant strain* The assumed displacement distributions should allow for all states of uniform strain (and stress) in an element. To achieve this requirement the second term in the polynomial expression must be linear. Note that if a sequence of approximate solutions is obtained using more elements for each solution, then these solutions should become closer and closer to any exact solution. For this so-called convergence to occur the strains in an element must approach a constant value. This requirement gives reliable modelling when an element becomes very small. To test that an element meets this requirement, the patch test has been developed; details of its application are given in texts listed in Sec. 3.10.

5. *Compatibility* The chosen displacement distributions should provide internal element compatibility and ideally maintain continuity of displacement (and rotation when bending is modelled) between elements. If all compatibility conditions are satisfied the element is said to be *conforming*. For plate bending and shell elements this requirement gives some difficulty, particularly as it has to be satisfied while also meeting requirement 1.

Convergence of a solution is guaranteed if the *completeness* requirements 2 to 4 are satisfied. The completeness of an element is assured if the polynomial expressions are of high enough degree and if no terms are omitted. Compatibility, requirement 5, must be met in the limit of element refinement and is met with fewer elements by most commonly used elements.

3.6 THE DIRECT APPROACH TO ELEMENT FORMULATION

3.6.1 A Bar Element

Figure 3.1 shows an element for a bar. For this one-dimensional element it will be assumed that it has constant properties along its length L. It is aligned with a local x-axis (and generally oriented with respect to the structure's global coordinate system) and its displacement can be defined in terms of the local axial displacement component u (i.e. parallel to the local x-axis). At each end of the element a node is placed giving it just two degrees of freedom, u_i and u_j, in the vector $\{\delta^e\}$, where subscripts i and j are used to indicate the element's left- and right-hand end nodes respectively. The element need not be given any real physical meaning here, other than representing a subregion of a physical bar.

It is now necessary to choose a mathematical form to represent the displacement over the element $u(x)$. A polynomial is the simplest form for the shape functions and, considering the requirements listed in Sec. 3.5.2, it is apparent that there is only one acceptable polynomial, namely

$$u(x) = u = \alpha_1 + \alpha_2 x = [\mathbf{f}]\{\alpha\} \tag{3.17}$$

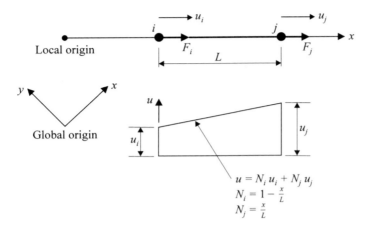

Figure 3.1 A two-noded linear bar element.

in which $[\mathbf{f}]$ is simply $[1 \ x]$ and the coefficients α_i are known as *generalized coefficients* which are dependent on the nodal displacements and coordinates. Expressions such as (3.17) are often referred to as *displacement functions*. Inserting the nodal boundary conditions of $u(x) = u_i$ at $x = 0$ and $u(x) = u_j$ at $x = L$ into (3.17) gives

$$\left\{ \begin{matrix} u_i \\ u_j \end{matrix} \right\} = \left[\begin{matrix} 1 & 0 \\ 1 & L \end{matrix} \right] \left\{ \begin{matrix} \alpha_1 \\ \alpha_2 \end{matrix} \right\} \qquad \text{or} \qquad \{\boldsymbol{\delta}^{\mathbf{e}}\} = [\mathbf{A}]\{\boldsymbol{\alpha}\} \qquad (3.18)$$

in which $[\mathbf{A}]$ is the Vandermode matrix. Combining (3.17) and (3.18) the shape function matrix $[\mathbf{N}]$ is defined from

$$\{\mathbf{u}\} = [\mathbf{f}][\mathbf{A}]^{-1}\{\boldsymbol{\delta}^{\mathbf{e}}\} = [\mathbf{N}]\{\boldsymbol{\delta}^{\mathbf{e}}\} \qquad (3.19)$$

Equation (3.17) is therefore $u = u_i + ((u_j - u_i)/L)x$ which, as shown in Fig. 3.1, is a linear variation in the displacement from u_i to u_j. $[\mathbf{N}]$ is the first of the three basic matrices on which the matrices $[\mathbf{k}]$ and $\{\mathbf{F}_{\mathbf{q}}^{\mathbf{e}}\}$ depend and for the bar element is $[1 - x/L \quad x/L]$. Note that $N_i = 1$ and $N_j = 0$ when $x = 0$ and that $N_i = 0$ and $N_j = 1$ when $x = L$, as required.

The second basic matrix is $[\mathbf{B}]$ and it is derived from (3.7) (a form of (2.9))

$$\{\boldsymbol{\varepsilon}\} = [\mathbf{B}]\{\boldsymbol{\delta}^{\mathbf{e}}\} = [\mathbf{f}'][\mathbf{A}]^{-1}\{\boldsymbol{\delta}^{\mathbf{e}}\} \qquad (3.20)$$

in which $\mathbf{f}'$ indicates differentiation of terms (here with respect to x). For the bar element this is

$$\varepsilon_x = \frac{\mathrm{d}u}{\mathrm{d}x} = \left[-\frac{1}{L} \quad \frac{1}{L} \right] \left\{ \begin{matrix} u_i \\ u_j \end{matrix} \right\}$$

in which the matrix $[\mathbf{B}]$ is $[-1/L \quad 1/L]$.

Finally, the third basic matrix [**D**] is derived from the one-dimensional stress–strain relationship (2.13). For the case where there is no initial strain or stress

$$\{\sigma\} = [\mathbf{D}]\{\varepsilon\} \qquad \text{or} \qquad \sigma_x = E\varepsilon_x \tag{3.21}$$

in which [**D**] has only one term E.

To complete the formulation of the matrix [**k**] using the integral in (3.9) it is noted that the terms in matrices [**B**] and [**D**] are constant such that the integration is simply

$$A \int_{x=0}^{L} \mathrm{d}x = AL \tag{3.22}$$

where A is the constant cross-sectional area. Now straightforward mathematical manipulation gives the 2 by 2 element matrix [**k**].

A similar approach enables the terms in the element consistent force vector $\{\mathbf{F_q^e}\}$ to be derived using (3.10) when one or more distributed loadings is present. Adding to the consistent nodal forces the concentrated nodal forces gives the element vector $\{\mathbf{F^e}\}$ and thus equation (1.1) ($\{\mathbf{F^e}\} = [\mathbf{k}]\{\boldsymbol{\delta^e}\}$) has the form

$$\begin{Bmatrix} F_i \\ F_j \end{Bmatrix} = \frac{EA}{L} \begin{bmatrix} 1 & -1 \\ -1 & 1 \end{bmatrix} \begin{Bmatrix} u_i \\ u_j \end{Bmatrix} \tag{3.23}$$

for a bar element with constant properties.

Building a mesh using bar elements, to model a bar with axial loading or a pin-jointed frame with loads concentrated at the joints, is carried out using the methods discussed in Chapter 5. However, as most of the elements are not positioned at the origin of the global axis system in the x-direction, transformations must be used to transform the local element into a global element before the system of equations is assembled. Note that an inclined bar element may have up to six degrees of freedom, as each node can have nonzero u-, v- and w-displacement components. Although a node may move in any global coordinate direction, a bar can carry only axial load, as the pin joint cannot transmit a moment, and thereby equation (2.23) is always valid.

Assembly of the algebraic equations from each element in (3.14) generates the global equations (1.2) (i.e. $\{\mathbf{F}\} = [\mathbf{K}]\{\boldsymbol{\delta}\}$). Methods of solution for the primary unknown nodal displacements ($\{\boldsymbol{\delta}\}$) are given in Sec. 3.9. Nodal displacements can then be extracted from the total set for the problem and, following appropriate transformation to the local axis system, the constant element stress (σ_x) determined by combining (3.20) and (3.21). It is worth remembering throughout the rest of the discussion on element formulation that the methods for assembly and solution of the algebraic equations are independent of the element formulation method, this being one of the most important advantages of the finite element method.

3.6.2 A Three-noded Triangular Element

A second example of the direct approach method illustrates how Turner *et al.* (1956) derived the characteristics for the three-noded triangular element shown in

Figs 1.2 and 1.3. They were interested in analysing aerospace structures consisting of thin flat plates with loads purely in the one plane (the x–y plane) and stress components uniform through the thickness. The element has straight sides and is assumed to have constant thickness. It has two displacement components u and v depending on the spatial global coordinates x and y. Such an element is often referred to in the literature as a plane stress or plane strain element. In order to satisfy requirements 1 to 5 of Sec. 3.5.2, the displacement functions are taken to be the complete linear polynomials

$$u = \alpha_1 + \alpha_2 x + \alpha_3 y$$
$$v = \alpha_4 + \alpha_5 x + \alpha_6 y$$

$$(3.24)$$

The same three basic matrices, as derived for the bar element Sec. 3.6.1, need to be obtained before the element characteristics can be obtained. A three-noded triangular element has six generalized coefficients in (3.24) and, therefore, six nodal degrees of freedom. To create matrix $[\mathbf{A}]$, now 6 by 6, the same procedure as discussed for the bar is used, by inserting into (3.24) the nodal displacements $(u_i, v_i, u_j, \ldots)$ and their corresponding global coordinates $(x_i, y_i, x_j, \ldots)$. The 3 by 3 $[\mathbf{D}]$ matrix, for an isotropic material, is obtained from either the plane stress or plane strain equations (2.29) or (2.30). Matrix $[\mathbf{B}]$ is obtained by using (3.24) with the strain–displacement relationships for the two-dimensional strain components ε_x, ε_y and γ_{xy}. Matrix $[\mathbf{B}]$ has, just as with the bar element, constant terms for an element of constant thickness, t. It follows from this that the integral over the volume will be simply tA (i.e. area $\times$ thickness) and so (3.9) has the simple form $[\mathbf{k}] = At[\mathbf{B}]^{\mathrm{T}}[\mathbf{D}][\mathbf{B}]$ in which $[\mathbf{B}] = [\mathbf{f}'(\mathbf{x}, \mathbf{y})][\mathbf{A}]^{-1}$. For this element it is not too difficult to derive explicitly the 6 by 6 matrix $[\mathbf{k}]$. The element's consistent force vector $\{\mathbf{F_q^e}\}$ is derived by following a similar procedure using (3.10).

Solving the simultaneous equations in (1.2) gives, for each element, six nodal displacements. These displacements, when inserted in (3.16) determine the element stresses (σ_x, σ_y, τ_{xy}), which are constant within the element and usually assigned to the centroid of the element. Alternatively, the stresses can be determined at the nodes by an averaging process of the values in adjoining elements.

One reason why this element has not been accepted as the universal element for irregular geometries is the development of the isoparametric element discussed in Sec. 3.7.1. Chapter 7 shows results for a plane stress example problem to illustrate the modelling ability of linear triangular elements.

3.6.3 Higher-order and Other Elements

The assumed simple displacement distributions for the bar (3.17) and the plane stress or plane strain (3.24) elements give a linear displacement variation and constant strain, and hence stress, in an element. Such an element is known as a *linear, simple* or *low-order element*. A *refined* or *higher-order element* is obtained by placing more nodes along the edges of an element and increasing the degree of the polynomial to account for the additional degrees of freedom. In the direct approach method the element edges are straight and the nodes have to be equally spaced. Such a procedure modifies a three-noded linear triangular element to a six-noded quadratic, nine-noded cubic or even higher-order elements.

The displacement functions for the six-noded triangular quadratic element use complete polynomials with those terms up to and including the quadratic terms. These are

$$u = \alpha_1 + \alpha_2 x + \alpha_3 y + \alpha_4 x^2 + \alpha_5 xy + \alpha_6 y^2$$
$$v = \alpha_7 + \alpha_8 x + \alpha_9 y + \alpha_{10} x^2 + \alpha_{11} xy + \alpha_{12} y^2$$

$$(3.25)$$

Here the displacement components vary quadratically and the strain (and stress) vary linearly within the element. It is therefore to be expected that this element gives better results where there are high strain (stress) gradients than does the three-noded element of equivalent size. Again, the three basic matrices $[\mathbf{N}]$, $[\mathbf{B}]$ and $[\mathbf{D}]$ are derived from which the 12 by 12 matrix $[\mathbf{k}]$ and the 6 by 1 vector $\{\mathbf{F_q^e}\}$ can be formulated.

Displacement functions (3.24) and (3.25) are *complete polynomials* (see requirements 1 to 4 in Sec. 3.5.2), which can be derived using *Pascal's triangle*. The form of the polynomials is such that there is geometric isotropy, i.e. the polynomials are balanced with respect to the x- and y-directions. If each side of the element has the same number of nodes the performance of these elements is independent of their orientation in the mesh and this is the reason why the element formulation is based on the global rather than a local coordinate system. All five requirements listed in Sec. 3.5.2 are satisfied by such displacement functions and convergence is therefore ensured. Hence these elements are conforming, or compatible. Triangular elements therefore have desirable properties and are a popular choice in analyses where the shape of the structure is irregular.

A family of two-dimensional straight-sided rectangular elements can also be formulated by the direct approach. Unlike the family of triangular elements their displacement functions are *incomplete polynomials* and so they do not give geometric isotropy. Because of their rectangular shape these elements are not as versatile as the equivalent triangular elements, as will be shown through examples in Chapters 7–9. They are, however, the preferred choice when modelling a structure with a topologically rectangular shape.

Finally, the matrix $[\mathbf{k}]$ and the vector $\{\mathbf{F_q^e}\}$ for three-dimensional eight-noded and twenty-noded solid brick elements and the four-noded tetrahedron (Table 3.2) can be derived by the direct approach and these elements have comparable properties to the linear and quadratic plane stress elements. Note that these matrices now have 24 and 60 terms associated with the number of degrees of freedom in these two elements.

From this discussion of the direct approach method it is apparent that the mathematical operations to formulate characteristics for two- and three-dimensional elements become increasingly more complicated as the order of the element (i.e. of the polynomial displacement functions) increases. Moreover, the inverse of matrix [**A**] may not exist. Even if it does, then the integration in (3.9) and (3.10) can often be very troublesome, requiring special mathematical procedures.

3.7 ISOPARAMETRIC ELEMENTS

Element types discussed so far have functions said to be of class C^0, meaning that the displacement components are continuous along interelement boundaries. Slopes, curvatures and higher-order derivatives of the displacements are not. The strain–displacement relationships and the stress–strain relationships, as given in Chapter 2, for C^0 elements show that strains and stresses depend on the first derivative (i.e. the rate of change of displacement components) and thus strains (and stresses) are discontinuous across interelement boundaries.

A second popular class of functions, C^1, for elements modelling bending deformation as shown in Table 3.2, will be discussed in Sec. 3.8.1. Here, the functions are those that have continuous displacement and slope along their interelement boundaries. These functions are used for beams, flat plate structures and general shell structures; this differs from the plane stress or plane strain simplification in that in-plane direct strains and direct stresses due to bending are assumed to vary linearly through the thickness of the structure.

The most popular and widely available of the C^0 elements, *isoparametric elements*, will be discussed.

3.7.1 Isoparametric Formulation of the Element Stiffness Matrix

There are several features of the isoparametric method that make it different from, and often more desirable than, the direct approach method. In the isoparametric method, a *master*, or parent, set of shape functions is first developed using the direct approach and then the master set is mapped or transformed onto each of the real elements in a mesh. Master elements are available for all of the elements in Tables 3.1 and 3.2. Their shape functions in matrix [**N**] of (3.5) are defined implicitly rather then explicitly such that there is now a greater reliance on numerical procedures to evaluate any integrals. Closed-form integration is possible in some special cases; but the expressions tend to be lengthy and so calculation is tedious and error-prone. However, the use of numerical integration schemes to integrate (3.9) and (3.10) for an element's characteristics can lead to different problems, and some of these will be presented.

For each element type the master element has a constant size and shape. For example, in two dimensions, triangular elements have the master element shown on the left in Fig. 3.2(a) and quadrilateral elements have the master element shown on the left in Fig. 3.2(b). Note that the master elements are defined in the natural coordinate system ξ and η. For the triangle it can be seen that it is placed in the range 0 to 1, whereas the quadrilateral master ranges from -1 to $+1$. Figure 3.2 also shows two curved-sided real elements, a six-node C^0-quadratic

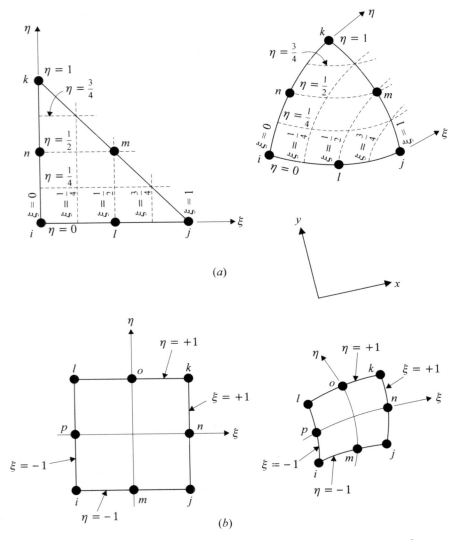

Figure 3.2 Isoparametric elements—master and real forms: (*a*) for six-noded C^0 quadratic triangular element, (*b*) for eight-noded C^0 quadratic quadrilateral element.

triangle and an eight-node C^0-quadratic quadrilateral. When transformed using appropriate shape functions these curved-sided elements in the global coordinate system become the appropriate master element in the natural coordinate system.

Note that 'parametric' refers to the use of the mapping and that 'iso', which means equal, refers to the fact that the mapping functions between global and natural coordinates are chosen to be the same as the shape functions used in (3.5). There are also C^0 sub- and super-parametric, as well as 'nonstandard' hierarchical elements which are used less frequently. Although C^1 elements for bending are not isoparametric, their element characteristics can be formulated using a master element and a natural coordinate system. Such elements have shape functions, for example, from the *Hermitian* family. They will not be discussed further here.

To explain how the isoparametric method is used to derive the stiffness matrix $[\mathbf{k}]$, the example of a bar element with constant properties will be repeated. This linear C^0 element has two nodes at each end. Its master element and a typical real element are shown in Fig. 3.3. The master element is placed in a natural coordinate system ξ such that node i is at $\xi = -1$ and node j is at $\xi = +1$. Hence the length of the master element is 2. Each real element will have some length L (defined as $x_j - x_i$) and any point x along its length is mapped from the master element by the expression

$$x = x(\xi) = \frac{1}{2}(1 - \xi)x_i + \frac{1}{2}(1 + \xi)x_j \qquad (3.26)$$

and so it can be seen that $N_i = \frac{1}{2}(1 - \xi)$ and $N_j = \frac{1}{2}(1 + \xi)$ which are the shape functions for the element.

The element displacement distribution can be defined by interpolating nodal displacements with the same shape functions, so

$$u = u(\xi) = \frac{1}{2}(1 - \xi)u_i + \frac{1}{2}(1 + \xi)u_j \qquad (3.27)$$

Now (3.27) can be written in the general form of (3.5) as

$$\{u\} = \left[\frac{1}{2}(1 - \xi) \quad \frac{1}{2}(1 + \xi)\right]\begin{Bmatrix} u_i \\ u_j \end{Bmatrix} \qquad \text{i.e. } \{\mathbf{u}\} = [\mathbf{N}]\{\boldsymbol{\delta}^e\}$$

The displacement u has the same linear form as that in (3.17), the starting point in the direct approach method, indicating that (3.27) satisfies the five requirements necessary for the element to provide convergence and compatibility given in Sec. 3.5.2. The element is seen to be isoparametric because the same shape function matrix $[\mathbf{N}]$ is used for interpolation of both *element geometry* and *element*

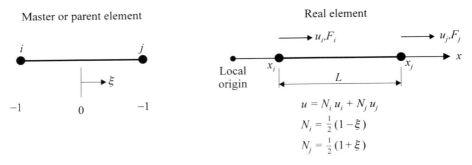

Figure 3.3 A two-noded C^0 linear isoparametric bar element.

displacements, i.e. when evaluating either the x-coordinate or the displacement u at any point along the master bar element the ξ-coordinate of that point is substituted into either (3.26) or (3.27).

Formulation of the stiffness matrix $[\mathbf{k}]$ requires that the matrix $[\mathbf{B}] = [\partial][\mathbf{N}]$ (see (2.10)) be known. Note that with the isoparametric method the basic shape function matrix $[\mathbf{N}]$ has been derived directly and, therefore, the need to derive and invert the Vandermode matrix $[\mathbf{A}]$ is removed, which is one of the major advantages of the method. The constant axial strain in the bar is, therefore, given by

$$\varepsilon_x = \frac{\mathrm{d}u}{\mathrm{d}x} = \left(\frac{\mathrm{d}}{\mathrm{d}x}[\mathbf{N}]\right)\begin{Bmatrix} u_i \\ u_j \end{Bmatrix} \quad \text{where} \quad \frac{\mathrm{d}}{\mathrm{d}x} = \frac{\mathrm{d}\xi}{\mathrm{d}x}\frac{\mathrm{d}}{\mathrm{d}\xi} \tag{3.28}$$

in which $[\partial]$ is the single term $\mathrm{d}/\mathrm{d}x$. To evaluate $\mathrm{d}/\mathrm{d}x[\mathbf{N}]$ the chain rule must be invoked because $[\mathbf{N}]$ is expressed in terms of ξ rather than in terms of x used to define strain and displacement. The term $\mathrm{d}\xi/\mathrm{d}x$ is not immediately available and first its *inverse* must be calculated from (3.26). For a one-dimensional element, the Jacobian J is defined as $\mathrm{d}x/\mathrm{d}\xi$ and so

$$J = \frac{\mathrm{d}}{\mathrm{d}\xi}[\mathbf{N}]\begin{Bmatrix} x_i \\ x_j \end{Bmatrix} = -\frac{1}{2}x_i + \frac{1}{2}x_j = \frac{L}{2} \tag{3.29}$$

as $x_j - x_i = L$. For the bar element shown in Fig. 3.3, J can be regarded as a scale factor that relates the physical length of the real bar element $\mathrm{d}x$ to corresponding length of the master bar element $\mathrm{d}\xi$; that is, $\mathrm{d}x = J\,\mathrm{d}\xi$.

Equation (3.9) for the bar element can now be written in two alternative forms as

$$[\mathbf{k}] = AE \int_{x_i}^{x_j} [\mathbf{B}]^T[\mathbf{B}]\mathrm{d}x = AE \int_{+1}^{-1} [\mathbf{B}]^T[\mathbf{B}]J\,\mathrm{d}\xi \tag{3.30}$$

For the simple bar element under consideration matrix $[\mathbf{B}]$ in (3.30) has the same constant terms in both integral forms as it is independent of both x and ξ. As a result numerical integration is not necessary for this element, a special case, and integration of (3.30) can be achieved explicitly to yield the matrix $[\mathbf{k}]$ in (3.23).

3.7.2 Lagrangian Elements

By placing nodes along the length of the straight bar element, a family of elements can be built up which are known as *Lagrangian elements* after the form of their shape functions. Figure 3.4 shows the next element in the series, the C^0-isoparametric quadratic bar element, and gives its shape functions. The element displacement, u, is now interpolated from three nodal values and (3.5) can be written as

$$u = \frac{1}{2}\xi(\xi - 1)u_i + (1 - \xi^2)u_j + \frac{1}{2}\xi(\xi + 1)u_k \tag{3.31}$$

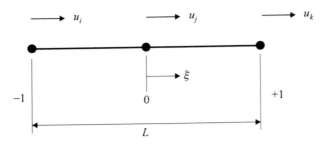

Shape functions $N_i = \frac{1}{2}\xi(\xi-1)$, $N_j = (1-\xi^2)$, $N_k = \frac{1}{2}\xi(\xi+1)$

Figure 3.4 A three-noded C^0 quadratic isoparametric bar element.

Note that node j is the interior node and that the shape functions are $N_i = \frac{1}{2}\xi(\xi - 1)$ and so on.

It is not necessary, as is the case with the direct approach method, to place the interior node at mid-length. If the third node is not at the centre, the mapping is nonuniform, i.e. the mapping is distorted, thus in the ξ-coordinate for the real element becomes compressed at one end and stretched at the other. To check that acceptable mapping occurs everywhere in the element it is required that the Jacobian $J > 0$. Note that if J, that is equal to $(L/2 + \xi(x_i - 2x_j + x_k))$, is negative the mapping is folding back on itself, which is clearly unacceptable. Hence, the interior node must always be located within $\frac{1}{4}L$ of the centre for the quadratic element shown in Fig. 3.4, but that in practice the best location is at the centre. If the third node is at a quarter point it produces a mathematical stress singularity and it is this feature of quadratic elements that allows the formulation of a cheap fracture mechanics element. For a cubic element, the standard location for the two interior nodes are at the $\frac{1}{3}$ points and there is a range either side of this in which J remains positive for the one-to-one mapping. It is common practice in commercial software to place interior nodes at locations known to provide the most well-behaved element and so the control of their positions is often outside the influence of the analyst.

3.7.3 Numerical Integration in One Dimension

With the procedure in Sec. 3.7.1, the element stiffness matrix for a quadratic element can be derived from (3.30) using (3.31) for the displacement u. As matrix $[\mathbf{B}]$ and J, if the interior node is not at the centre, are now functions of ξ, it is necessary to evaluate the terms of the 3 by 3 matrix $[\mathbf{k}]$ by numerical integration.

The process of numerical integration transforms an analytical integral into the sum of a finite number of terms. Note the similarity with the philosophy of the general finite element method where a continuum has been discretized into the sum of a finite number of elements (or subregions). Hence, for a one-dimensional integral such as that in (3.30), with n sampling points

$$\int_{x_i}^{x_j} f(x)\, dx = \int_{-1}^{+1} I(\xi)\, d\xi \equiv \sum_{l=1}^{n} w_{nl} I(\xi_{nl}) \tag{3.32}$$

Here, l is the number of the sampling points being considered, w_{nl} are weighted factors and ξ_{nl} are points at which the integrand is evaluated, called *quadrature points, integration points, sampling points* or *Gauss points*. In the finite element formulation $I(\xi)$ represents the terms, depending on element type, from the matrix product $[\mathbf{B}]^T[\mathbf{D}][\mathbf{B}]J$.

To evaluate (3.32) there are many well-established quadrature formulae available. Examples of these include the Newton–Cotes rules, such as the well-known Simpson's rule, or the trapezoidal rule in which the sampling points are equally spaced. Experience has shown, however, that Gauss quadrature rules are more precise than these. For n sampling points in (3.32), the Gauss rules integrate exactly a polynomial of degree $2n - 1$. When the limits of the integration are -1 to $+1$ the rule is the *Gauss–Legendre rule*, although, as will be seen shortly, this is not restricted to one dimension.

In Gauss–Legendre quadrature, the positions of the sampling points ξ_{nl} are the zeros of the nth-degree Legendre polynomial and are often known as Gauss points. Both weights w_{nl} and the Gauss points ξ_{nl} have numerical values that are given in Table 3.1 for values of n up to 3. Note that the Gauss points are located symmetrically with respect to the centre of the integration interval ($\xi = 0$) and that symmetrically equivalent points have the same weights, e.g. when n is 2 $w_{22} = w_{21} = 1$. Table 3.3 gives also the order of accuracy as a power of the element length $h \equiv L$.

To approximate integral (3.32) in the simplest way only one sampling point can be used, i.e. $n = 1$. From Table 3.3 this gives $\xi_{11} = 0$ and $w_{11} = 2$ to give (3.32) as

$$\int_{-1}^{+1} I(\xi)\, d\xi \equiv w_{11} I(\xi_{11}) = 2I(0) \tag{3.33}$$

Table 3.3 Gauss points and associated weights for Gauss–Legendre quadrature

Number of Gauss points, n	Accuracy of quadrature	Gauss points, ξ_{nl}	Weights, w_{nl}
1	$O(h^2)$	$\xi_{11} = 1$	$w_{11} = 2$
2	$O(h^4)$	$\xi_{21} = -\xi_{22} = -0.577\,35$	$w_{11} = w_{22} = 1$
3	$O(h^6)$	$\xi_{31} = -\xi_{33} = -0.774\,60$	$w_{31} = w_{33} = 0.555\,55$
		$\xi_{32} = 0$	$w_{32} = 0.888\,88$

in which $I(0)$ is the value of the integrand at the midpoint and 2 is the length of the interval. Thus the area under a curve $I(\xi)$ is approximated by a rectangular area and this is exact if curve $I(\xi)$ describes a straight line of finite slope. If the integrand is not of this form then a better approximation for the integral is to take $n = 2$ to give the integral as

$$\int_{-1}^{+1} I(\xi)\,\mathrm{d}\xi \equiv w_{21}I(\xi_{12}) + w_{22}I(\xi_{22}) = I(-0.577\,35) + I(0.577\,35) \qquad (3.34)$$

This two-point rule integrates a cubic polynomial exactly. Note that when deriving the terms in matrix $[\mathbf{k}]$ using (3.32) the matrix terms $I(\xi)$ have a constant denominator if the terms in $[\mathbf{B}]$ are constant (e.g. for the bar elements in Figs 3.3 and 3.4 (if the third node is at the centre). Then and only then is numerical integration exact. With matrix $[\mathbf{B}]$ depending on $1/J$ this can be true only when J is constant. If, however, the element is distorted from its master shape the terms in $I(\xi)$ are the ratio of two polynomials in ξ and numerical integration is generally inexact, regardless of the rule used. In other words, the ratio of two polynomials is not a polynomial in general. Distortion of an element from its parent shape always introduces discretization errors into the finite element analysis. This aspect of modelling has been taken into account when developing the examples for Chapters 7 to 9.

Clearly, the number of Gauss points can be increased until acceptable accuracy is obtained. However, there will be an optimum rule and therefore the question of the best order of quadrature must be considered. There is no simple answer, although one-dimensional elements are much less complicated to evaluate than two- and three-dimensional elements. According to Burnett (1987) many of the quadrature rules commonly used in commercial codes for a given element type have evolved more from numerical experience than from theoretical arguments, the latter being developed to explain the success or failure of the former.

3.7.4 Serendipity Elements and the Comparison with Lagrangian Elements

It might seem that two-dimensional elements could have a whole host of shapes such as general polygons or conic sections (circles and ellipses). However, of all possible geometries, only three-sided and four-sided polygons (with straight or curved sides) seem to be well suited to application of the isoparametric method. Families of C^0-isoparametric triangular and quadrilateral elements have been created following similar arguments to those used in the direct approach method. Both these element shapes have a family of elements based on Lagrangian shape functions. There is also a second family of quadrilateral elements which do not have interior nodes and whose shape functions are known as *serendipity shape functions*. Here, 'serendipity' means that the shape functions have been determined by inspection. The very high order Lagrangian and serendipity elements, say quintic or above, at present are mostly of academic interest, as commercial packages usually allow elements only up to the order of cubic elements.

Shape functions for these element types are given in many texts and so only the C^0-isoparametric serendipity shape functions for the quadratic eight-noded element in Fig. 3.2(b) are reproduced here. They are

$$N_i = \frac{1}{4}(1 - \xi)(1 - \eta)(-\xi - \eta - 1)$$

$$N_j = \frac{1}{4}(1 + \xi)(1 - \eta)(\xi - \eta - 1)$$

$$N_k = \frac{1}{4}(1 + \xi)(1 + \eta)(\xi + \eta - 1)$$

$$N_l = \frac{1}{4}(1 - \xi)(1 + \eta)(-\xi + \eta - 1)$$

$$N_m = \frac{1}{2}(1 - \xi^2)(1 - \eta) \qquad\qquad (3.35)$$

$$N_n = \frac{1}{2}(1 + \xi)(1 - \eta^2)$$

$$N_o = \frac{1}{2}(1 - \xi^2)(1 + \eta)$$

$$N_p = \frac{1}{2}(1 - \xi)(1 - \eta^2)$$

The majority of software packages provide two- and three-dimensional elements which do not have interior nodes. There is some disagreement between authors on the merits of Lagrangian and serendipity elements and on the usefulness of interior nodes. Burnett (1987) considers Lagrangian quadrilateral elements to have two disadvantages; that the degree of polynomial completeness is low for the order of the element and that the presence of interior nodes can create problems when generating a mesh. Zienkiewicz and Taylor (1989) also state that the Lagrangian family is limited not only because of the large number of internal nodes but also because of their poorer curve-fitting properties. For these reasons they advocate the use of the serendipity family instead. In contrast, Cook *et al.* (1989) state that the Lagrangian shape functions, because they 'do not leave out the middle-part' of Pascal's triangle, allow better accuracy than is found using the equivalent serendipity shape functions. Furthermore, the nine-noded Lagrangian quadrilateral element is much less sensitive than the eight-noded serendipity quadrilateral element to distortion of its geometry and to placing 'midside' nodes away from the centre. When consulting the user manuals for commercial software, the analyst often finds that the method used to formulate the equations for an element is not given and this makes assessment of the code's performance more difficult. Such conflicting guidance in the standard texts illustrates the importance of the activities of NAFEMS and, in particular, their benchmarks.

3.7.5 Numerical Integration in Two Dimensions

Clearly, it must be possible to integrate functions over two and three dimensions. This is a simple extension of the work over one dimension given in Sec. 3.7.3 and so only the two-dimensional version is given here. Over two dimensions, numerical integration is given by

$$\int \int f(x, y)\, dx\, dy = \int_{-1}^{+1} \int_{-1}^{+1} I(\xi, \eta) d\xi\, d\eta \equiv \sum_{k=1}^{n} \sum_{l=1}^{n} w_{nk} w_{nl} I(\xi_{nk}, \eta_{nl}) \qquad (3.36)$$

where n Gauss points are used in each of the two coordinate directions, k and l are the indicators for the sampling point being considered and $I(\xi, \eta)$ is the integrand from the stiffness matrix $[\mathbf{k}]$ which is again obtained from the matrix multiplication $[\mathbf{B}]^T [\mathbf{D}][\mathbf{B}]\, J$. Previously it was noted that the Jacobian J for a one-dimensional element is just a single term, $d/d\xi = [\mathbf{J}]d/dx$. In two dimensions J becomes the determinant of the 2 by 2 Jacobian matrix

$$[\mathbf{J}(\xi, \eta)] = \begin{bmatrix} \dfrac{\partial x}{\partial \xi} & \dfrac{\partial y}{\partial \xi} \\ \dfrac{\partial x}{\partial \eta} & \dfrac{\partial y}{\partial \eta} \end{bmatrix} \qquad (3.37)$$

Equation (3.36) can be used with the Gauss–Legendre rule for a quadrilateral element as the integration limits are from -1 to $+1$. Now, the sampling points form a two-dimensional array and the weight applied at each Gauss point can be seen in (3.36) to be a product of the one-dimensional weights given in Table 3.3. Table 3.4 gives values for the Gauss points ξ_{nk} and η_{nl} and the weights $w_{nk} \times w_{nl}$ for values of n up to 3.

Triangular elements, because of to their shape, cannot have integration limits of -1 to $+1$ as shown in Fig. 3.2(a) and so the Gauss–Legendre rule is inappropriate. Consequently, another Gaussian integration scheme has to be available to derive the matrix $[\mathbf{k}]$ and the vector $\{\mathbf{F_q^e}\}$ for the Lagrangian family of triangular elements.

3.7.6 Accuracy of Isoparametric Elements

Now a number of the features of a quadrilateral element and numerical integration which relate to the accuracy of calculation of the stiffness matrix $[\mathbf{k}]$ will be discussed. Note that these features are similar when considering a triangular element. As usual, the criterion for an acceptable element is that the mapping is one-to-one, which is achieved as long as J is positive and not zero. To achieve this, two restrictions must be applied to the shape of an element. For an eight-noded quadratic element as shown in Fig. 3.5, the interior corners must have an included angle of less than 180° and the midside nodes must be placed within $\pm L/4$ of the centre position on a side. Note here the similarity to the restrictions on a quadratic bar element discussed previously. It has also been found, through

Table 3.4 Gauss points and associated weights for two-dimensional product-type Gauss–Legendre quadrature

n	Number of Gauss points $n \times n$	Accuracy of quadrature	Gauss points ξ_{nk}, η_{nl}	Weights $w_{nk} \times w_{nl}$
1	1 (1×1)	$O(h^2)$		$4(= 2 \times 2)$ at centre
2	4 (2×2)	$O(h^4)$	$\xi = -1\sqrt{3}$ $\xi = +1/\sqrt{3}$ $\eta = +1/\sqrt{3}$ $\eta = -1/\sqrt{3}$	$1(= 1 \times 1)$ at points 1,2,3,4
3	9 (3×3)	$O(h^6)$	$\xi = -\sqrt{3}/\sqrt{5}$ $\xi = +\sqrt{3}/\sqrt{5}$ $\eta = +\sqrt{3}/\sqrt{5}$ $\eta = -\sqrt{3}/\sqrt{5}$	$25/81(= 5/9 \times 5/9)$ at points 1,3,7,9 $40/81(= 5/9 \times 8/9)$ at points 2,4,6,8 $64/81(= 8/9 \times 8/9)$ at point 5

numerical experiments, that accuracy is seriously impaired when the curvature of a side is so great that the circular arc defined by the three nodes of a side subtends an angle of 180° or greater. This is also shown in Fig. 3.5.

In finite element models the most frequently used element shape has straight sides with the nodes equally spaced along the sides of the element. For example, quadratic elements, in this case, have their midside nodes placed centrally. Also, when the shape of a quadrilateral element is a parallelogram it can be shown that the J is simply one-quarter of the area of the element, that there is no mapping distortion and that numerical integration using (3.36) can be exact. The next most frequently used shape has three straight sides and one curved side. In this case, where there is some distortion of the element, J is nonzero and acceptable, provided that the midside nodes on the curved side remain at their preferred locations or are within the prescribed limits as shown in Fig. 3.5.

Only infrequently does an element have more than one curved side and this is usually at a corner of a structure. The ability of an element's shape to model exactly the shape of the surface of a structure depends on the description of the surface geometry itself. If the surface is flat the lowest-order element to fit exactly is a linear element. However, if the surface has a parabolic shape then a quadratic

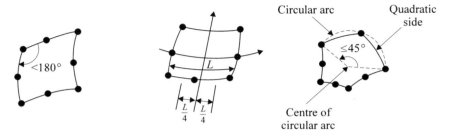

Figure 3.5 Some limitations on C^0 quadratic isoparametric elements for mapping problems.

element is sufficient. In this case, the use of linear elements may provide a poor approximation to the surface geometry unless many such elements are used. In fact, whenever the geometry of a surface is described by a polynomial of higher order than an element's displacement distributions (i.e. the shape functions for isoparametric elements) modelling inaccuracy is incurred. In modern CAD packages the order of a polynomial can be around 20 and so there is often a modelling error when a structure is discretized into elements.

Distortion can also exist if the shape of an element is an extremely narrow parallelogram. The stiffness matrix [**k**] for such an element is ill-conditioned because two of its interior angles are close to 180°. In addition, the area of the element and hence its J tend to zero. This leads the element to have a much greater stiffness than a less distorted element and, in turn, is a further source of ill-conditioning leading to round-off errors when solving (1.2) by the techniques discussed in Sec. 3.9.

If follows from this discussion that, when creating a mesh, elements that are not too dissimilar from their master shape (Fig. 3.2) are preferred. The following guidelines are often applied:

- The aspect ratio of an element should not be greater than 5 to 1 (or 10 to 1).
- Interior corner angles should be within 20° or 30° of a right angle. If this condition can not be achieved then a triangular element would be better.

In an attempt to maintain solution accuracy, some commercial packages enforce a strict limit on the maximum and minimum allowable values for the interior corner angles. Beyond this limit the solution will not proceed. Further, a second more generous limit is often set beyond which a warning is given to alert the analyst to the fact that element performance in terms of accuracy is being loss. Acceptable element shapes can also be enforced during the automatic creation of an unstructured mesh (see Chapter 5).

Extension of the procedures outlined here to formulate the stiffness matrix [**k**] for three-dimensional solid brick elements (eight-noded linear and twenty-noded quadratic) does not pose any further fundamental difficulties. Brick elements usually have no interior node and their shape functions belong to the serendipity family. Other shapes of elements that are commonly available include the tetrahedron and wedge.

As alluded to when introducing numerical integration, it is often unclear which quadrature rule should be used when forming the stiffness matrix [**k**]. Now, *full integration* is defined as taking place when a quadrature rule is used that is sufficient to provide the exact integral of all terms if the element is undistorted. Note that the same full integration does not exactly integrate all the terms if the sides are curved, for the reasons already given above. Full integration is the only sure way to avoid pitfalls such as mesh instability, where the elements deform in such a way that the strains at the Gauss points are zero and hence these do not contribute to the strain energy in (3.9). Further details on mesh instabilities (which have nothing to do with structural buckling) and the various methods to control them are given in standard texts.

A lower-order quadrature rule, called *reduced integration*, may be desirable for two reasons:

1. Since the expense of generating matrix [**k**] by numerical integration is proportional to the number of Gauss points used (i.e. to the value of n), using lower-order quadrature leads to lower computational cost.
2. A lower-order rule tends to soften the stiffness of an element, thus compensating for the over-stiff behaviour inherent with an assumed displacement distribution. Softening comes about because certain of the higher-order polynomial terms vanish at the Gauss points so that they make no contribution to strain energy. This has the additional benefit of reducing the values of stresses that are overestimated by displacement elements.

Experience has shown that for an isoparametric element the best quadrature rule is usually the lowest that computes volume correctly and does not produce mesh instabilities. For example, a Gauss rule with order $n = 2$ (reduced integration) is favoured for eight-noded quadratic quadrilateral elements and for eight-noded linear brick elements. There are, therefore, four and eight Gauss points, respectively. Note that all numerical integration rules are valid only in the limit of the element becoming small and hence a loss in accuracy must be incurred when elements are large.

3.7.7 Isoparametric Formulation of the Element Consistent Force Vector

Throughout the discussion of the formulation of element characteristics there has been a concentration on the element stiffness matrix [**k**], virtually ignoring the element consistent force vector $\{\mathbf{F_q^e}\}$. Most of the features discussed with regard to formulating [**k**] are true when deriving $\{\mathbf{F_q^e}\}$. As explained in Sec. 3.1.3, the term 'consistent' follows from the fact that the distributed loading is transformed into a set of nodal forces that are consistent with the shape functions in matrix [**N**] of (3.5). It is found that when the sides of an element are straight and any edge nodes are at their preferred locations then the force vector is exact in terms of the external work done by the distributed loading.

Let us now consider an eight-noded isoparametric quadrilateral element, whose shape functions are given in (3.35), having one external edge with a surface pressure applied to it. To simplify the explanation of how to derive $\{\mathbf{F_q^e}\}$, the element is assumed to be rectangular with sides of length a and b, to have constant thickness, to have the midnodes at the midpoints and to have the loaded edge parallel to the ξ and global x-coordinates (i.e. on the side with nodes i, m and j). The applied pressure is constant and has the value q_y with units of $\mathrm{N\,mm^{-1}}$ per unit thickness. Body forces, initial stresses and initial strains are assumed to be zero. Figure 3.6(a) shows the situation described.

Recalling expression (3.12) for $\{\mathbf{F_q^e}\}$ and noting that the only nonzero distributed load term in the surface traction vector $\{\bar{\mathbf{X}}\}$ is $\bar{Y}(=q_y)$, it is now

$$\{\mathbf{F_q^e}\} = \int_{-1}^{+1} [\mathbf{N}(\xi, -1)]^{\mathrm{T}} \begin{Bmatrix} 0 \\ q_y \end{Bmatrix} J_\Gamma(\xi, -1)\mathrm{d}\xi = \sum_{l=1}^{l} w_{nl}\left([\mathbf{N}(\xi_n, -1)]^{\mathrm{T}} \begin{Bmatrix} 0 \\ q_y \end{Bmatrix} J_\Gamma(\xi_n, -1)\right)$$

(3.38)

in which $J_\Gamma(\xi, \eta)$ is known as the *boundary Jacobian* since it is derived from the ratio of differential arc lengths (Burnett, 1987). For the simple case in Fig. 3.6(a), $J_\Gamma(\xi, -1)$ is the constant $a/2$. Equation (3.38) is readily solved either directly or by numerical integration (using the rules in Table 3.3) to yield the following nodal force terms:

$$F_{yi} = \frac{1}{6}aq_y \qquad F_{ym} = \frac{4}{6}aq_y \qquad F_{yi} = \frac{1}{6}aq_y$$

(3.39)

The vector $\{\mathbf{F_q^e}\}$ consists of 16 terms for this element as there are 2 degrees of freedom per node, but all the other 13 terms here are zero. It can be shown that the consistent element forces are those which, if applied in the opposite sense as

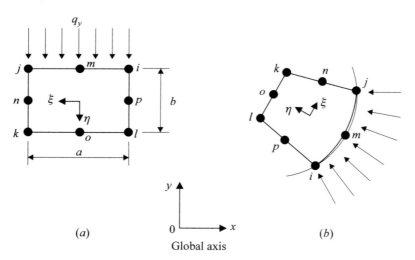

(a)   (b)

Global axis

Figure 3.6 Evaluation of element consistent force terms for quadrilateral elements with an edge pressure: (a) on a straight edge, (b) on a curved edge.

constraints, would keep all the nodal displacements zero in the presence of the true loading. Note that by making the nodal forces consistent, the midside node attracts four times the force value of the corner nodes and not twice the value if the total load aq_y is apportioned between the nodes on the basis of a lumped surface area. If the latter method is used it is known as *lumping* or *inconsistent loading* (Cook et al., 1989). The same result is obtained if the loaded side is one of the three straight sides of a six-noded triangular element.

If the third node is moved away from the centre or if the element has a curved side, as shown in Fig. 3.6(b), the boundary Jacobian $J_\Gamma(\xi, \eta)$ and $\bar{X}$ and $\bar{Y}$ are no longer constant (Burnett, 1987). These factors and any lack of element fit to the surface geometry of a structure (Fig. 3.6b) provide a different set of nodal forces (with F_{xi}, F_{xm}, F_{xj} also nonzero if the side is curved). Such nodal forces, although consistent with the shape functions of the element, do not provide the exact external work. It can be seen that, as when formulating [**k**], any distortion to the master element causes a loss in the accuracy of $\{\mathbf{F_q^e}\}$ and hence in the finite element solution. A similar procedure is followed to determine the nodal forces due to other distributed loads. Those nodal forces due to body forces, initial stresses and initial strains are derived from volume integrals so that the procedures discussed when deriving [**k**] are applicable.

The above argument for the development of $\{\mathbf{F_q^e}\}$ can be extended to generalized shell and flat plate bending elements with a loading in the form of a face (or surface) distributed pressure.

3.7.8 Calculating Stresses with Isoparametric Elements

Once [**k**] and $\{\mathbf{F_q^e}\}$ for an element have been derived, and then calculated for each element in a mesh, it is not difficult to create the global system of equations, as defined by $\{\mathbf{F}\} = [\mathbf{K}]\{\boldsymbol{\delta}\}$, Eq. (3.14), that describes the full structural problem. Before solving these simultaneous equations for the nodal displacements $\{\boldsymbol{\delta}\}$ it is necessary to eliminate, by a *condensation technique* described by Cook et al. (1989), those equations associated with the interior nodal degrees of freedom of higher-order Lagrangian elements. The equations generated using isoparametric elements which do not have interior nodes require no such treatment and their element characteristics are dealt with in the same way as those formulated by the direct approach method.

Having solved (1.2), the nodal displacements $\{\boldsymbol{\delta}^e\}$ of each element are extracted from $\{\boldsymbol{\delta}\}$ and inserted into (3.16) to evaluate the stresses within an element. If initial values are neglected, then the stresses are given by

$$\{\boldsymbol{\sigma}\} = [\mathbf{D}]\,[\mathbf{B}]\,\{\boldsymbol{\delta}^e\}$$

in which the stresses (e.g. σ_x, σ_y and so on) are a function of the global coordinates and the [**B**] matrix is in terms of natural coordinates. This combination of coordinate systems poses the question of where in the element the stresses should be calculated. For isoparametric elements, stresses (especially shear stresses) are most accurate at the Gauss points of a quadrature rule one order lower than that for full integration of [**k**] (Burnett, 1987; Cook *et al.*, 1989). Because stresses are dependent on strains and these are derived from differentials of the element displacements, it may be expected that stresses will be less accurate than the primary unknowns, the nodal displacements, and this is true for elements formulated using the direct approach method. However, under certain conditions the stresses for isoparametric elements are not less accurate than displacements. These stresses are *superaccurate* or *superconvergent* at the Gauss points because they have there the same degree of accuracy as displacements. Nodal stresses are evaluated by interpolating and/or extrapolating the stresses at Gauss points.

Whatever method is used to formulate an element's characteristics, the calculation of stresses is always relative to a rectilinear coordinate system and often the system is Cartesian. As discussed in Sec. 2.2.6 the acceptance of a structural design can be aided by the use of an appropriate failure criterion such as von Mises' criterion. Such criteria, be they stress or strain based, are functions of the principal values (Sec. 2.2.2 and 2.2.4) and these are obtained by transformation from the Cartesian values. It will be seen in Chapters 6–9 how element stresses are manipulated to produce contour plots, this being the most popular method used to display results.

3.8 OTHER TYPES OF ELEMENTS

3.8.1 Elements for Bending

These one- and two-dimensional elements are used to model beams and flat plate structures in bending by assuming that in-plane direct strains vary linearly through the thickness of the structure and that there is no axial strain component. However, bending and in-plane elements can be combined to produce flat shell elements. These are not considered to be the best shell elements.

Now it is found that both displacement and slope components have to be continuous along interelement boundaries, while curvatures and higher-order derivatives are not continuous. To illustrate the features of this class of element the three basic matrices in (3.9) and (3.10) will be derived for a pure bending beam element (i.e. neglecting shear deformation due to through-thickness shearing). This one-dimensional element (see Table 3.2) can have the same form as the bar element in Fig. 3.1 which has two end nodes. In this development it is assumed that the element has constant flexural rigidity EI_a (I_a is the second moment of area about the axis of bending) along its length L. Its centroidal axis is aligned with the local x-axis and the y-axis is vertically downwards. Each end node has two degrees of freedom, one transverse deflection and one slope (or rotation), giving four degrees of freedom (i.e., v_i, $\theta_i = dv/dx_{(x=i)}$, v_j and $\theta_j = dv/dx_{(x=j)}$), in vector $\{\boldsymbol{\delta}^e\}$.

Enforcement of requirements 1 to 5 in Sec. 3.5.2 defines the displacement function to be

$$v = \alpha_1 + \alpha_2 x + \alpha_3 x^2 + \alpha_4 x^3 \qquad (3.40)$$

from which matrix $[\mathbf{A}]$ (and $[\mathbf{A}]^{-1}$) is derived by inserting into expression (3.40) the nodal coordinates and nodal displacements.

Derivation of matrices $[\mathbf{B}]$ and $[\mathbf{D}]$ is different to that for class C^0 elements as it is convenient to have the stress–strain relationship (3.1) without initial values, in the form of the well-known *moment–curvature* expression

$$M(x) = -EI_a \frac{d^2 v}{dx^2} \qquad \text{i.e. } \{\boldsymbol{\sigma}\} = [\mathbf{D}]\{\boldsymbol{\varepsilon}\} \qquad (3.41)$$

which shows that when bending is being modelled the terms in the strain matrix $\{\boldsymbol{\varepsilon}\}$ are *curvatures* and the terms in the stress matrix $\{\boldsymbol{\sigma}\}$ are *bending moments*. Note that the strain is given by $\varepsilon_x = \pm y \, d^2 v/dx^2$. Matrix $[\mathbf{D}]$ for the beam element has the single term EI_a and matrix $[\mathbf{B}]$ is derived from the second derivative of displacement function (3.41) by following the procedure given for the bar element. The exact element stiffness matrix can then be readily derived from (3.9) as the integration can be done explicitly. The form of $[\mathbf{k}]$ for the beam element is identical to the *slope–deflection* equations used in the direct stiffness method.

After Eq. (1.2) for a whole beam problem have been solved, the four element nodal displacements are used to evaluate the bending moment anywhere along the element. The direct stress that determines the resistance of the beam is then calculated from the well-known bending relationship (2.25), i.e., $M/I_a = \sigma_x/\pm y$, where the sign convention is that positive moment M causes the beam to sag.

Extension to two-dimensional flat plate bending elements (Fig. 3.1) follows a similar procedure to that discussed for two-dimensional plane stress and plane strain elements. However, it not practical to have a displacement function for the vertical displacement, $w(x, y)$, that is compatible and this leads to *nonconforming elements*. Some plate elements have been tuned to remedy problems due to shear and locking. In practice these elements are found to work well providing sufficient elements are used to minimize the error due to the lack of interelement compatibility. However, as with the case of shell elements, no outright champion has yet been formulated.

3.8.2   Special Elements

The scope of this text allows detailed discussion of only the basic features of the linear elastic static finite element method. There are many other features which need attention as they represent further evidence of the diversity and versatility of the method. The purpose of this section is to convey a number of these features and to direct the reader to references giving further details. In addition to the references given here, information on these features, when they are available, can be found in the user manuals of commercial analysis software.

With reference to Tables 3.1 and 3.2 it is found that a number of element types have not yet been dealt with. Of these types there are the one-dimensional elements to model a rigid link, a linear spring, a beam with offset, a tapered beam and a Mindlin shear deformable beam (i.e. for deep beams).

Axisymmetric solid and axisymmetric shell element types (both thick and thin) are often used in problem-solving and a major application for these elements is in the analysis of pressure vessels and tanks. Axisymmetric elements are for problems concerning the stress distribution in structures of revolution under axisymmetric loading. By symmetry, an element for the solid of revolution is two-dimensional and the formulation of its element characteristics is analogous to that for plane stress and plane strain elements.

Shell elements can be in two or three dimensions. They are used to model structures with curved surfaces in space that have loading acting normal to the surface. The properties of a shell are lumped to act at the midsurface of the shell to develop the mathematical theory from which element characteristics are derived in the usual way. Often a shell is thin in comparison with its span and then classical thin shell theory is applicable. However, when it is necessary to include in the analysis the effects of transverse shear deformation and perhaps of through-thickness direct stress, the theory is that of a thick shell. Note that the Mindlin plate theory allows transverse displacement and rotations to be independent such that the element formulation only requires C^0 continuity. This has a major advantage of simplifying the shape functions in (3.5) needed to formulate the element characteristics. General shell elements display bending and membrane (in-plane) deformations, with the latter dominating deformation if the structure is to be effective. It is readily seen that flat plate elements (in-plane and bending) are in fact degenerate forms of a generalized thin shell element. Some packages therefore do not provide separate flat plate elements and expect the analyst to use the available shell elements to model such cases.

Two-dimensional classical shell theory produces equations that are difficult to solve. The equations in terms of displacements are often complicated unless many approximations are made (Calladine, 1983). There is no consensus as to which approximations are acceptable, so it is found that there a number of shell theories (e.g. Donnell (1933), Flügge (1973) and Vlasov (1964)). All these theories are limited to small displacements. To formulate element characteristics it is helpful if displacement distributions are chosen, Eq. (3.5), that will satisfy all five requirements presented in Sec. 3.5.2—for shell elements this is difficult, if not impossible. Elements for shells are therefore the most difficult elements to derive, and because of the various shell theories and approaches to the problem there are a number of elements available. Problems with three- and four-sided elements based on thin shell analysis have led to a move away from classical shell formulation and towards treating shell elements as a special version of the three-dimensional solid element, when the through-thickness dimension becomes relatively small. These quadratic isoparametric elements that simulate behaviour of the Mindlin shell element are not without their problems. For example, they do not pass the patch test and so one of the functions of NAFEMS is to publicize the performance of shell elements either singly or in groups. They are, however, favoured by the analysis community.

Elements described so far have sides with the same number of nodes. Packages do offer so-called *transition elements* where, for example, one side of a quadratic quadrilateral element does not have a third node and therefore, along that edge, behaves as a linear element. This is achieved by forcing the midside node to lie on a straight line between corner nodes using constraint equations so that no gap can open up and continuity is maintained. Transition elements are used in mesh construction when the element order changes. A situation when this may be desirable is where, because strain (and stress) gradients are high, regions in a structure need quadratic elements for an accurate solution, but where in the adjacent regions linear elements are acceptable as the gradients there are much lower. By integrating into a mesh both linear and quadratic elements the numbers of degrees of freedom, and so the round-off errors and the computational cost of solving the problem, can be minimized for a given accuracy.

Special element types exist to model cracks in a fracture mechanics analysis and to model an infinite media such as the ground supporting a structure. Other specialized elements are available which do not represent a continuum but which have their place when modelling practical problems. Such elements include the lumped mass, gap or interface elements, damper elements, spring elements of zero length and constraint elements.

Finite element packages offer a range of material properties for an element type. Here, for convenience, isotropic materials (Sec. 2.2.6) have been chosen to show how the material's elastic constants (i.e., E and ν) in matrix $[\mathbf{D}]$ are used in the formulation of an element stiffness matrix terms and in the calculation of the element stresses. Some packages allow all 36 C_{ij} terms (see Eq. (2.15)) in $[\mathbf{D}]$ to be independently defined such that any anisotropic material can be described. A majority of packages are less flexible but, because of their importance in structural applications, have the facility to model laminated materials. Laminates often have a sheet structure and consist of multi-layers of a fibre-reinforced material (Halpin, 1992), and it is therefore plate and shell elements (thin and thick) that have this material option. For a general laminate, the properties attached to a general shell element are formulated using lamination theory, details of which are to be found in texts on fibre-reinforced composite materials.

When developing the finite element method equations in Sec. 3.2.2 it was stated that initial strains and stresses can be generated because of the thermal expansion property of a material. It is often the situation that a load-bearing structure has residual stresses at room temperature after cooling down from the processing temperature, e.g. hot rolled steel, or has to operate at a temperature other than room temperature, and for the finite element analysis to be accurate it is imperative to account for any thermal effects. Finite element packages therefore offer a range of thermal elements to solve heat transfer problems. To determine the overall loading for a structural analysis the results from such an analysis are used to establish the thermally generated loadings that are added to the other loading types (see (3.10) and (3.12)).

In Chapter 10, both buckling and dynamic analysis will be discussed when they are performed using the linear small-displacement elements developed in this section. This will be followed by the introduction of the finite element method in analysing problems that are time-dependent and nonlinear in terms of either geometry or their material properties.

3.9 PRODUCING A SOLUTION FROM THE ELEMENT EQUATIONS

In the previous sections of this chapter the ways in which numerical analogues of the governing partial differential equations can be produced have been shown. For all types of elements, equations are produced for each element which relate deformation of the whole element to the discrete displacements and forces at the nodes of the element. These equations know nothing of the behaviour of the full structure and so, now, there must be a discussion as to how these equations are combined, a process known as assembly, and then processed such that the boundary conditions of the problem are applied before a static solution for the full structure is found.

3.9.1 Assembly into a Global Matrix

On each element of the structure, an equation or equations can be developed to describe the behaviour over the element. For example:

$$[\mathbf{k}]\{\boldsymbol{\delta}^e\} = \{\mathbf{F}^e\}$$

which is (1.1) again. Here, $[\mathbf{k}]$ is the stiffness matrix for the element, $\{\boldsymbol{\delta}^e\}$ is the vector of nodal displacements and $\{\mathbf{F}^e\}$ is the vector of forces at the nodes. Such equations are always singular and so they have to be combined together to form a global set of matrix equations and then the boundary conditions for the whole structure can be applied to remove any singularities. This works only when the structure has no mechanisms.

Figure 3.7 shows a simple mesh of two-dimensional plane stress triangular elements with two degrees of freedom per node. The mesh consists of three triangular elements and five nodes in total. Element equations can be written for each element in the form

$$\begin{bmatrix} a_{11}^e & a_{12}^e & a_{13}^e & a_{14}^e & a_{15}^e & a_{16}^e \\ a_{21}^e & a_{22}^e & a_{23}^e & a_{24}^e & a_{25}^e & a_{26}^e \\ a_{31}^e & a_{32}^e & a_{33}^e & a_{34}^e & a_{35}^e & a_{36}^e \\ a_{41}^e & a_{42}^e & a_{43}^e & a_{44}^e & a_{45}^e & a_{46}^e \\ a_{51}^e & a_{52}^e & a_{53}^e & a_{54}^e & a_{55}^e & a_{56}^e \\ a_{61}^e & a_{62}^e & a_{63}^e & a_{64}^e & a_{65}^e & a_{66}^e \end{bmatrix} \begin{bmatrix} u_i^e \\ v_i^e \\ u_j^e \\ v_j^e \\ u_k^e \\ v_k^e \end{bmatrix} = \begin{bmatrix} F_{ix}^e \\ F_{iy}^e \\ F_{jx}^e \\ F_{jy}^e \\ F_{kx}^e \\ F_{ky}^e \end{bmatrix} \tag{3.42}$$

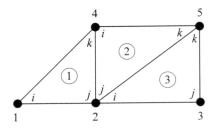

Figure 3.7 A simple mesh of triangular elements.

where the superscript e refers to the number of the element and the other notation is the same as that used before and in Fig. 3.7.

These element equations can be modified, for each element, to relate to the 10 degrees of freedom global problem by using the global node numbers shown in Fig. 3.7. For example, local node k on element 1 is global node 4, local node k on element 2 is global node 5 and so on. These expanded element equations are:

$$
\begin{bmatrix}
a_{11}^1 & a_{12}^1 & a_{13}^1 & a_{14}^1 & 0 & 0 & a_{17}^1 & a_{18}^1 & 0 & 0 \\
a_{21}^1 & a_{22}^1 & a_{23}^1 & a_{24}^1 & 0 & 0 & a_{27}^1 & a_{28}^1 & 0 & 0 \\
a_{31}^1 & a_{32}^1 & a_{33}^1 & a_{34}^1 & 0 & 0 & a_{37}^1 & a_{38}^1 & 0 & 0 \\
a_{41}^1 & a_{42}^1 & a_{43}^1 & a_{44}^1 & 0 & 0 & a_{47}^1 & a_{48}^1 & 0 & 0 \\
0 & 0 & 0 & 0 & 0 & 0 & 0 & 0 & 0 & 0 \\
0 & 0 & 0 & 0 & 0 & 0 & 0 & 0 & 0 & 0 \\
a_{71}^1 & a_{72}^1 & a_{73}^1 & a_{74}^1 & 0 & 0 & a_{77}^1 & a_{78}^1 & 0 & 0 \\
a_{81}^1 & a_{82}^1 & a_{83}^1 & a_{84}^1 & 0 & 0 & a_{87}^1 & a_{88}^1 & 0 & 0 \\
0 & 0 & 0 & 0 & 0 & 0 & 0 & 0 & 0 & 0 \\
0 & 0 & 0 & 0 & 0 & 0 & 0 & 0 & 0 & 0
\end{bmatrix}
\begin{bmatrix} u_1 \\ v_1 \\ u_2 \\ v_2 \\ u_3 \\ v_3 \\ u_4 \\ v_4 \\ u_5 \\ v_5 \end{bmatrix}
=
\begin{bmatrix} F_{1x}^1 \\ F_{1y}^1 \\ F_{2x}^1 \\ F_{2y}^1 \\ 0 \\ 0 \\ F_{4x}^1 \\ F_{4y}^1 \\ 0 \\ 0 \end{bmatrix}
$$

$$
\begin{bmatrix}
0 & 0 & 0 & 0 & 0 & 0 & 0 & 0 & 0 & 0 \\
0 & 0 & 0 & 0 & 0 & 0 & 0 & 0 & 0 & 0 \\
0 & 0 & a_{33}^2 & a_{34}^2 & 0 & 0 & a_{37}^2 & a_{38}^2 & a_{39}^2 & a_{310}^2 \\
0 & 0 & a_{43}^2 & a_{44}^2 & 0 & 0 & a_{47}^2 & a_{48}^2 & a_{49}^2 & a_{410}^2 \\
0 & 0 & 0 & 0 & 0 & 0 & 0 & 0 & 0 & 0 \\
0 & 0 & 0 & 0 & 0 & 0 & 0 & 0 & 0 & 0 \\
0 & 0 & a_{73}^2 & a_{74}^2 & 0 & 0 & a_{77}^2 & a_{78}^2 & a_{79}^2 & a_{710}^2 \\
0 & 0 & a_{83}^2 & a_{84}^2 & 0 & 0 & a_{87}^2 & a_{88}^2 & a_{89}^3 & a_{810}^2 \\
0 & 0 & a_{93}^2 & a_{94}^2 & 0 & 0 & a_{97}^2 & a_{98}^2 & a_{99}^3 & a_{910}^2 \\
0 & 0 & a_{103}^2 & a_{104}^2 & 0 & 0 & a_{107}^2 & a_{108}^2 & a_{109}^2 & a_{1010}^2
\end{bmatrix}
\begin{bmatrix} u_1 \\ v_1 \\ u_2 \\ v_2 \\ u_3 \\ v_3 \\ u_4 \\ v_4 \\ u_5 \\ v_5 \end{bmatrix}
=
\begin{bmatrix} 0 \\ 0 \\ F_{2x}^2 \\ F_{2y}^2 \\ 0 \\ 0 \\ F_{4x}^2 \\ F_{4y}^2 \\ F_{5x}^2 \\ F_{5y}^2 \end{bmatrix}
\tag{3.43}
$$

$$
\begin{bmatrix}
0 & 0 & 0 & 0 & 0 & 0 & 0 & 0 & 0 & 0 \\
0 & 0 & 0 & 0 & 0 & 0 & 0 & 0 & 0 & 0 \\
0 & 0 & a_{33}^3 & a_{34}^3 & a_{35}^3 & a_{36}^3 & 0 & 0 & a_{39}^3 & a_{310}^3 \\
0 & 0 & a_{43}^3 & a_{44}^3 & a_{45}^3 & a_{46}^3 & 0 & 0 & a_{49}^3 & a_{410}^3 \\
0 & 0 & a_{53}^3 & a_{54}^3 & a_{55}^3 & a_{56}^3 & 0 & 0 & a_{59}^3 & a_{510}^3 \\
0 & 0 & a_{63}^3 & a_{64}^3 & a_{65}^3 & a_{66}^3 & 0 & 0 & a_{69}^3 & a_{610}^3 \\
0 & 0 & 0 & 0 & 0 & 0 & 0 & 0 & 0 & 0 \\
0 & 0 & 0 & 0 & 0 & 0 & 0 & 0 & 0 & 0 \\
0 & 0 & a_{93}^3 & a_{94}^3 & a_{95}^3 & a_{96}^3 & 0 & 0 & a_{99}^3 & a_{910}^3 \\
0 & 0 & a_{103}^3 & a_{104}^3 & a_{105}^3 & a_{106}^3 & 0 & 0 & a_{109}^3 & a_{1010}^3
\end{bmatrix}
\begin{bmatrix} u_1 \\ v_1 \\ u_2 \\ v_2 \\ u_3 \\ v_3 \\ u_4 \\ v_4 \\ u_5 \\ v_5 \end{bmatrix}
=
\begin{bmatrix} 0 \\ 0 \\ F_{2x}^3 \\ F_{2y}^3 \\ F_{3x}^3 \\ F_{3y}^3 \\ 0 \\ 0 \\ F_{5x}^3 \\ F_{5y}^3 \end{bmatrix}
$$

Note that now the degrees of freedom are u and v at each of the five nodes and that each element has zeros in both the left-hand side matrix and the right-hand side vector where the position corresponds to a global node that is not attached to the element in question. These equations are then summed to give the final equation, which is

$$
\begin{bmatrix}
a^1_{11} & a^1_{12} & a^1_{13} & a^1_{14} & 0 & 0 & a^1_{17} & a^1_{18} & 0 & 0 \\
a^1_{21} & a^1_{22} & a^1_{23} & a^1_{24} & 0 & 0 & a^1_{27} & a^1_{28} & 0 & 0 \\
a^1_{31} & a^1_{32} & \begin{matrix}a^1_{33}+a^2_{33}\\+a^3_{33}\end{matrix} & \begin{matrix}a^1_{34}+a^2_{34}\\+a^3_{34}\end{matrix} & a^3_{35} & a^3_{36} & a^1_{37}+a^2_{37} & a^1_{38}+a^2_{38} & a^2_{39}+a^3_{39} & a^2_{310}+a^3_{310} \\
a^1_{41} & a^1_{42} & \begin{matrix}a^1_{43}+a^2_{43}\\+a^3_{43}\end{matrix} & \begin{matrix}a^1_{44}+a^2_{44}\\+a^3_{44}\end{matrix} & a^3_{45} & a^3_{46} & a^1_{47}+a^2_{47} & a^1_{48}+a^2_{48} & a^2_{49}+a^3_{49} & a^2_{410}+a^3_{410} \\
0 & 0 & a^3_{53} & a^3_{54} & a^3_{55} & a^3_{56} & 0 & 0 & a^3_{59} & a^3_{510} \\
0 & 0 & a^3_{63} & a^3_{64} & a^3_{65} & a^3_{66} & 0 & 0 & a^3_{69} & a^3_{610} \\
a^1_{71} & a^1_{72} & a^1_{73}+a^2_{73} & a^1_{74}+a^2_{74} & 0 & 0 & a^1_{77}+a^2_{77} & a^1_{78}+a^2_{78} & a^2_{79} & a^2_{710} \\
a^1_{81} & a^1_{82} & a^1_{83}+a^2_{83} & a^1_{84}+a^2_{84} & 0 & 0 & a^1_{87}+a^2_{87} & a^1_{88}+a^2_{88} & a^2_{89} & a^2_{810} \\
0 & 0 & a^2_{93}+a^3_{93} & a^2_{94}+a^3_{94} & a^3_{95} & a^3_{96} & a^2_{97} & a^2_{98} & a^2_{99}+a^3_{99} & a^2_{910}+a^3_{910} \\
0 & 0 & a^2_{103}+a^3_{103} & a^2_{104}+a^3_{104} & a^3_{105} & a^3_{106} & a^2_{107} & a^2_{108} & a^2_{109}+a^3_{109} & a^2_{1010}+a^3_{1010}
\end{bmatrix}
\begin{bmatrix}
u_1 \\ v_1 \\ u_2 \\ v_2 \\ u_3 \\ v_3 \\ u_4 \\ v_4 \\ u_5 \\ v_5
\end{bmatrix}
=
\begin{bmatrix}
F_{1x} \\ F_{1y} \\ F_{2x} \\ F_{2y} \\ F_{3x} \\ F_{3y} \\ F_{4x} \\ F_{4y} \\ F_{5x} \\ F_{5y}
\end{bmatrix}
$$

$$(3.44)$$

To carry out this process all that is required is for the element assembly software to know, for each element, the global node number of each of the nodes attached to it. This is the connectivity list. Simple algorithms can then be used to loop over each element in the mesh, placing the elements of the local element equations into the global left-hand side matrix and the global right-hand side vector. This gives the governing equations in the form of (1.2). Note that (3.44) is sparse even for this simple example. The zeros in the global stiffness matrix come about naturally during the assembly process, confirming the fact that certain nodes have no direct effect on other nodes. For example, global node 3 has no effect on global nodes 1 and 4.

3.9.2 Applying Boundary Conditions

At this stage in the process, having completed the formation of the global matrices, the boundary conditions of the problem can be applied. Typically this involves the specification of applied nodal forces and displacement restraints. Figure 3.8 shows the same mesh as Fig. 3.7 with the addition of boundary conditions. Here, it can be seen that at nodes 1 and 2 the v-displacement is zero, that at node 3 both the u- and

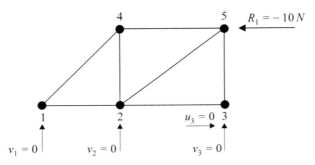

Figure 3.8 Boundary conditions for the simple mesh.

v-displacements are zero and that at node 5 there is an applied concentrated force of 10 N in the negative x-direction. To impose the displacement constraints v_1, v_2, v_3 and u_3 must be set to zero. This is forced on the matrix system by setting all values of the row and column associated with these degrees of freedom in the left-hand matrix to zero except for the diagonal which is set to one. Then the right-hand vector has to be modified. To do this (3.11) is used, where the element right-hand vector is made up of the contribution due to the applied external loads and any internal forces as defined by (3.10). In this case all the internal forces are zero and the only applied load is at the fifth node in the negative x-direction. Hence the right-hand vector becomes all zero except for the applied load, and then the appropriate values for the restraints, which are zero in this case, multiplied by the column values that have been zeroed from the left-hand side matrix are subtracted.

This gives

$$
\begin{bmatrix}
a_{11} & 0 & a_{13} & 0 & 0 & 0 & a_{17} & a_{18} & 0 & 0 \\
0 & 1 & 0 & 0 & 0 & 0 & 0 & 0 & 0 & 0 \\
a_{31} & 0 & a_{33} & 0 & 0 & 0 & a_{37} & a_{38} & a_{39} & a_{310} \\
0 & 0 & 0 & 1 & 0 & 0 & 0 & 0 & 0 & 0 \\
0 & 0 & 0 & 0 & 1 & 0 & 0 & 0 & 0 & 0 \\
0 & 0 & 0 & 0 & 0 & 1 & 0 & 0 & 0 & 0 \\
a_{71} & 0 & a_{73} & 0 & 0 & 0 & a_{77} & a_{78} & a_{79} & a_{710} \\
a_{81} & 0 & a_{83} & 0 & 0 & 0 & a_{87} & a_{88} & a_{89} & a_{810} \\
0 & 0 & a_{93} & 0 & 0 & 0 & a_{97} & a_{98} & a_{99} & a_{910} \\
0 & 0 & a_{103} & 0 & 0 & 0 & a_{107} & a_{108} & a_{109} & a_{1010}
\end{bmatrix}
\begin{bmatrix}
u_1 \\ v_1 \\ u_2 \\ v_2 \\ u_3 \\ v_3 \\ u_4 \\ v_4 \\ u_5 \\ v_5
\end{bmatrix}
=
\begin{bmatrix}
0 - a_{12}\cdot 0 - a_{14}\cdot 0 \\
0 \\
0 - a_{32}\cdot 0 - a_{34}\cdot 0 - a_{35}\cdot 0 - a_{36}\cdot 0 \\
0 \\
0 \\
0 \\
0 - a_{72}\cdot 0 - a_{74}\cdot 0 \\
0 - a_{82}\cdot 0 - a_{84}\cdot 0 \\
R - a_{94}\cdot 0 - a_{95}\cdot 0 - a_{96}\cdot 0 \\
0 - a_{104}\cdot 0 - a_{105}\cdot 0 - a_{106}\cdot 0
\end{bmatrix}
$$

$$(3.45)$$

As an aside, note that this process has mimicked the mathematical imposition of boundary conditions in that where the displacement is restrained the effect of any applied loads is overwritten.

3.9.3 Solving the Simultaneous Equations

The global equations must now be solved. For linear static problems without any time variation, this means solving just one set of linear simultaneous equations. Note that this is not the situation if gap elements are present as an iterative procedure is necessary. There are many ways of solving the equations, and each solver will have its own way, or ways, of finding a solution from the equations. For real structural problems the number of degrees of freedom is typically of the order of thousands or tens of thousands and so robust direct methods are used to find solutions. However, as problems become more complex so the computational effort needed to find a solution starts to rise as a proportion of the time taken to set up the element equations. When the solution of the equations consumes a large amount of computational effort, there are great benefits to be gained from using iterative methods to solve the simultaneous equations.

Solving any set of simultaneous equations is the process of finding a vector $\{\mathbf{x}\}$ that satisfies the general matrix equation

$$[\mathbf{A}]\{\mathbf{x}\} = \{\mathbf{b}\} \tag{3.46}$$

For finite element problems, (3.46) is directly equivalent to (1.2). To find a solution, the inverse of the matrix $[\mathbf{A}]$ must be determined and then both sides of (3.46) are premultiplied by this inverse to give

$$\{\mathbf{x}\} = [\mathbf{A}]^{-1}\{\mathbf{b}\} \qquad (3.47)$$

When the number of equations is relatively small then the inverse of the matrix $[\mathbf{A}]$ can be found using *direct methods*. Typically, a version of the method called LU decomposition is used. However, to reduce the requirements on computer memory the frontal method as described by Lewis and Ward (1991) and Taylor and Hughes (1981) has been developed. This combines the process of assembly and solution, handling the matrices element after element. It will not be discussed further here.

In LU decomposition the matrix $[\mathbf{A}]$ is decomposed into two other matrices:

$$[\mathbf{A}] = [\mathbf{L}][\mathbf{U}] \qquad (3.48)$$

where $[\mathbf{L}]$ is a lower triangular matrix and $[\mathbf{U}]$ is an upper triangular matrix. Once the matrix $[\mathbf{A}]$ has been decomposed into $[\mathbf{L}]$ and $[\mathbf{U}]$ the solution is easy to find. This is because the simple triangular structure makes it easy to find the inverses of these matrices. Depending on the symmetry properties of the matrix $[\mathbf{A}]$, the method has a number of variants, such as the Crout method, for nonsymmetric matrices and the Cholesky method for symmetric matrices.

As a rule of thumb, the time taken for solution using LU decomposition is proportional to the cube of the number of unknowns for a fully populated matrix, or to the product of the number of unknowns and the half-bandwidth squared for a sparse matrix. In some texts the terms 'half-bandwidth' and 'bandwidth' are synonymous. As was seen for the simple problem shown in Fig. 3.7, some of the terms in the matrix are zero because of the structure of the mesh. For example, node 1 has no direct connection to node 3 or node 5 and the matrix structure reflects this. For meshes of realistic problems many values in the matrix are zero, and the half-bandwidth is the maximum distance between the diagonal and the first nonzero element of a column in the upper triangle of the left-hand-side matrix. As the matrices are usually symmetrical, and even nonsymmetric ones have a symmetrical structure, this is the same as the maximum distance between the diagonal and the first nonzero element of a row in the lower triangle. Figure 3.9 shows this for a typical matrix.

Owing to the nature of the solution effort, if a matrix is very large then direct methods take a long time to produce a solution. In cases of complex structural analysis with nonlinearity and time variance, such large matrices are often produced. To reduce the time taken *iterative methods* are usually used. With these methods some guess at the solution vector $\{\mathbf{x}\}$ is made and then updated using the vector $\{\mathbf{x}\}$ and the coefficients of the matrix $[\mathbf{A}]$ and vector $\{\mathbf{b}\}$. In fact, many iterative schemes can be used, so to illustrate the use of these methods let us consider the solution of (3.46). For a system of three equations:

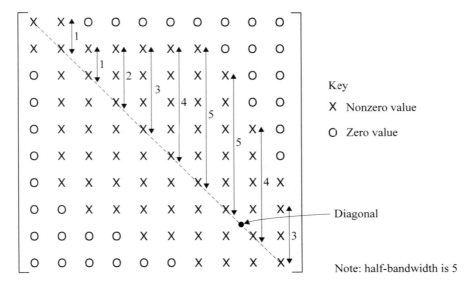

Figure 3.9 The half-bandwidth of a sparse matrix.

$$a_{11}x_1 + a_{12}x_2 + a_{13}x_3 = b_1$$
$$a_{21}x_1 + a_{22}x_2 + a_{23}x_3 = b_2 \qquad (3.49)$$
$$a_{31}x_1 + a_{32}x_2 + a_{33}x_3 = b_3$$

The simplest iterative schemes are those due to Jacobi and Gauss–Seidel. In both of these methods the equations are transformed to give

$$x_1 = \frac{1}{a_{11}}(b_1 - a_{12}x_2 - a_{13}x_3)$$
$$x_2 = \frac{1}{a_{22}}(b_2 - a_{21}x_1 - a_{23}x_3) \qquad (3.50)$$
$$x_3 = \frac{1}{a_{33}}(b_3 - a_{31}x_1 - a_{32}x_2)$$

Now it can be seen that the terms on the diagonal of $[\mathbf{A}]$, i.e. a_{ii}, must not be zero for these methods to work. Fortunately, zero values are not generated in static analysis using the finite element method. In the Jacobi method, the right-hand side of (3.50) is taken to be the known values at the nth iteration and the left-hand side to be the new values at the $n + 1$th iteration. This gives

$$x_1^{n+1} = \frac{1}{a_{11}}(b_1 - a_{12}x_2^n - a_{13}x_3^n)$$
$$x_2^{n+1} = \frac{1}{a_{22}}(b_2 - a_{21}x_1^n - a_{23}x_3^n) \qquad (3.51)$$
$$x_3^{n+1} = \frac{1}{a_{33}}(b_3 - a_{31}x_1^n - a_{32}x_2^n)$$

whereas the Gauss–Seidel method continually updates the right-hand side where it can to give

$$x_1^{n+1} = \frac{1}{a_{11}}(b_1 - a_{12}x_2^n - a_{13}x_3^n)$$

$$x_2^{n+1} = \frac{1}{a_{22}}(b_2 - a_{21}x_1^{n+1} - a_{23}x_3^n)$$

$$x_3^{n+1} = \frac{1}{a_{33}}(b_3 - a_{31}x_1^{n+1} - a_{32}x_2^{n+1})$$

(3.52)

The next level of complexity involves point relaxation methods. These methods operate on the error in the solution vector $\{\mathbf{x}\}$ which is known as the *residual error* and is defined as

$$\{\mathbf{r}\} = \{\mathbf{b}\} - [\mathbf{A}]\{\mathbf{x}\}$$

(3.53)

This should become smaller from iteration to iteration. To modify the previous method (3.52) is taken and x_i^n is both added to and subtracted from the right-hand side, to give

$$x_1^{n+1} = x_1^n + \left[\frac{1}{a_{11}}(b_1 - a_{11}x_1^n - a_{12}x_2^n - a_{13}x_3^n)\right]$$

$$x_2^{n+1} = x_2^n + \left[\frac{1}{a_{22}}(b_2 - a_{21}x_1^{n+1} - a_{22}x_2^n - a_{23}x_3^n)\right]$$

$$x_3^{n+1} = x_3^n + \left[\frac{1}{a_{33}}(b_3 - a_{31}x_1^{n+1} - a_{32}x_2^{n+1} - a_{33}x_3^n)\right]$$

(3.54)

Now the expressions in square brackets are the terms of the residual $\{\mathbf{r}\}$, and as these should tend to zero as the process proceeds, attempts can be made to accelerate the process by multiplying the right-hand side by a *relaxation factor*, ω, to give

$$x_1^{n+1} = x_1^n + \left[\frac{\omega}{a_{11}}(b_1 - a_{11}x_1^n - a_{12}x_2^n - a_{13}x_3^n)\right]$$

$$x_2^{n+1} = x_2^n + \left[\frac{\omega}{a_{22}}(b_2 - a_{21}x_1^{n+1} - a_{22}x_2^n - a_{23}x_3^n)\right]$$

$$x_3^{n+1} = x_3^n + \left[\frac{\omega}{a_{33}}(b_3 - a_{31}x_1^{n+1} - a_{32}x_2^{n+1} - a_{33}x_3^n)\right]$$

(3.55)

For most systems of equations the value of ω can be set to a value somewhere between 1 and 2. In this case the method is the *successive overrelaxation method*. If ω is unity then the method becomes the original Gauss–Seidel method.

Line relaxation methods where subsystems of the equations are used to provide an update to a number of values at one time can also be used. Also, more advanced methods are continuously being developed, as further research is carried out. Such research is needed to reduce the computational effort required to solve large systems of equations on supercomputers. Advanced methods include Stone's

(Smith, 1985) strongly implicit procedure, preconditioning methods and conjugate gradient methods which can be seen as matrix manipulation procedures, and multigrid methods which calculate the solution on a series of coarse and fine grids in space, swapping between the grids in such a way that any errors are smoothed out.

3.10 REFERENCES AND FURTHER READING

Some of the key texts that cover the mathematics of the finite element method are Burnett (1987), Cook *et al.* (1989), Stasa (1986) and Zienkiewicz and Taylor (1989). Of these texts, that by Burnett provides the most rigorous and detailed discussion of the mathematical aspects of the finite element method for solid mechanics. Other general works are Cook (1995), NAFEMS (1986) and Hughes and Hinton (1986). The work of Taig is mentioned by Robinson (1985), although no work was published on Taig's method until Irons (1966). For the Rayleigh–Ritz method see Ritz (1909) and Reddy (1984).

Numerical solution methods are described by the authors above as well as by Smith (1985) and Hoffman (1992).

4

IMPLEMENTING A COMPUTER-BASED
ANALYSIS PROCEDURE

This chapter contains material previously published in Shaw (1992), by kind permission of Prentice Hall.

In Chapters 1–3, the design process, the evaluation of designs and the use of computers in evaluation were considered, together with the techniques for analysing structural problems exactly (analytically) and also by numerical approximation. Now is the time to review the information in Chapters 1–3 and formulate a procedure for using computers as an aid to the structural analyst. First, the key stages of the structural analysis process must be defined, if only in overview; within subsequent chapters these stages will be described in detail. Then the types of computer software that are available will be discussed so that appropriate software can be used to produce the relevant information during the analysis procedure. This will be followed by a discussion of computer hardware architectures and the problem types that can be solved with them. Finally, the ways in which software and hardware are acquired, together with the necessary human skills, will be outlined.

4.1 A PROCESS FOR ANALYSING A STRUCTURE

In Chapter 2, a mathematical analysis of stress and strain led to a series of partial differential equations that govern all structural behaviour. In Chapter 3 functional forms of these partial differential equations were discretized on an element of the structure to produce a numerical analogue of the equations on each element. When boundary conditions and, possibly, initial conditions that define the

91

structural problem being considered have been applied, the equations can be solved to produce a numerical simulation of the problem.

Regardless of whether locally written or commercial finite element software is being used, the analyst must provide a common set of information to the software before a simulation is possible. From Chapters 2 and 3, this information must include the following:

- A mesh of nodes and elements, developed from the physical geometry being considered, on which the equations can be developed and the variables stored.
- Boundary conditions for loads and displacements (restraints) so that the problem is correctly defined.
- Possibly initial conditions. These might define the initial state of the structure for a transient problem or define the internal forces in the structure due to some initial stress or strain field, or due to body forces such as gravity.
- Material properties such as Young's modulus and Poisson's ratio.
- Control parameters that influence the solution of the equations and the ways in which the results are stored.

For an analyst the process that is followed must first of all generate the above information. Then the process must allow the analyst to check that the results are usable before the relevant engineering information is extracted from the model. One form of the full process can be divided into a series of stages as shown in Fig. 4.1:

- *Initial thinking*   Quite often the time taken for an analysis can be minimized if the analyst makes a detailed study of the problem to be solved before touching a computer. This should ensure that the correct problem is solved and useful results are produced. In this initial thinking stage the analyst needs to consider the structural problem and understand as much as possible about it. Here, sources of the geometrical shape of a design, its expected structural behaviour when subjected to loads, typical applied loads and material properties are particularly important. To gather this information liaison is necessary with a variety of people such as design engineers, process engineers, materials specialists and technicians. Further information can also be gained from a literature search of previous work in the area.
- *Mesh generation*   This second stage involves the production of a computer model of the geometry of the design being considered, if such a model does not exist already. Then the geometry is discretized to form a mesh of elements and nodes. This means that the structure is broken down into small subregions on which the numerical analogue of the governing equations can be developed.
- *Specification of the numerical problem*   In the initial thinking stage, relevant boundary conditions will have been determined in terms of both the physics of the problem and the physical location of the shape. Now, however, nodes and elements have been created in the mesh generation stage and so these boundary conditions must be reinterpreted in terms of the mesh. This will involve the specification of, for example, the forces to be applied at certain nodes, the

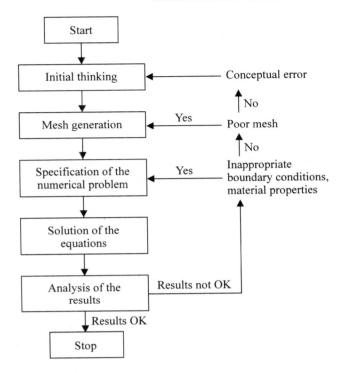

Figure 4.1 The analysis process.

pressures at certain element faces and the known displacements at given nodes. Similarly, initial conditions may also have to be specified in more complex calculations. Finally, the material properties of all the elements will need to be defined.

- *Solution of the equations* Having done all of the above, the solution software can be run. This will calculate a numerical solution for the mesh being considered with the appropriate boundary conditions and material properties.

- *Analysis of the results* Finally, when results have been generated, the analyst must first check to see that the numerical solution is satisfactory before determining the required engineering data from the solution.

While this series of stages might be performed one after the other in a linear sequence, it has already been seen in Chapter 1 that many design activities involve some iteration. This part of the design process is no exception as there are many potential sources of error that might lead to the engineering results being of little use. To reduce the possibilities for errors to occur, a mixture of sensible user experience and good management practice must be demonstrated during the analysis. Here, user experience includes common sense and the development of the art of analysis, with an analyst making suitable choices and approximations during the analysis. Equally, good management practice includes careful documentation of the work as well as constant checking of progress.

Suitable common checking procedures will be described in the subsequent chapters. If these procedures show that the analysis is not progressing satisfactorily then it may be necessary to repeat some of the stages. Yet again, iteration is necessary if the final solution is to be improved. However, by using computer technology, the refinement of the computational model is usually straightforward.

Common interactions between the stages are shown in Fig. 4.1. Note that internal checking within a given stage has not been shown here, but rather that some typical paths between stages are illustrated. From this, the iterative nature of the process becomes clear, with apparent problems forcing the analyst to loop back to some previous stage in the process in an attempt to improve the solution produced.

4.2 GENERIC SOFTWARE PACKAGES FOR STRESS ANALYSIS

Having looked at the stages of the analysis process, the software that is required to assist the analyst in carrying out the tasks that form each of these stages can now be considered. Taking each of the stages in turn, it can be seen that the first stage requires no computing, the second and third require input from the analyst, the fourth involves computing but no actual input from the analyst and the fifth stage requires the analyst to evaluate the results. From this an appropriate set of software might consist of

- a pre-processor
- a solver
- a post-processor.

Three separate programs could be used, but in most cases the extensive graphical operations required for both pre- and post-processors, coupled with the high levels of user input, lead many systems to have a combined pre- and post-processor system. Also, as much data has sometimes to be transferred from one format to another or from one machine type to another, a variety of utility programs are also used. These relationships are shown in Fig. 4.2.

4.2.1 Pre-processors

All of the tasks that take place before the numerical solution process are called pre-processing. As stated above, the second and third phases of the analysis process involve computing and so the pre-processor software usually assists the analyst in carrying out the following operations:

- definition of geometry in computational form
- definition of a mesh of nodes and elements to represent the geometry

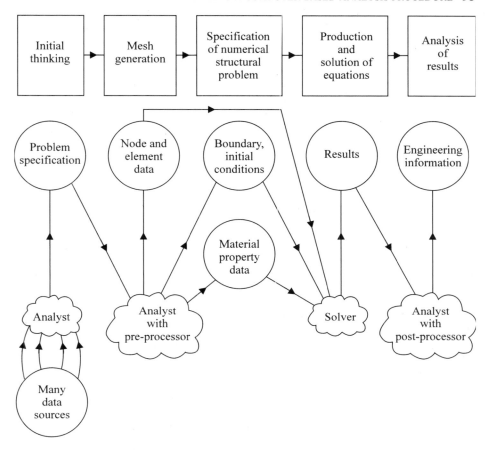

Figure 4.2 Interactions between the analyst and software during the analysis.

- definition of appropriate sections of boundaries of the geometry, in terms of the mesh data, at which boundary conditions will be applied
- application of the boundary conditions
- application, where necessary, of the initial conditions
- definition of material and physical properties for groups of elements
- application of control parameters for the solver.

As all these tasks involve the user interacting with the computer, so pre- processing programs tend to have a user-friendly, graphical interface, probably using the so-called Windows, Icons, Menus, Pull-down Screens (or WIMPS) technology. Using such an interface, coupled with high-quality visual display systems, allows various parameters to be set and the resulting changes to be seen quickly. This is of particular importance when the the geometry of the design is being created and when the mesh is being built. Extensive use of colour is made to assist the analyst.

In many cases, similar designs will be analysed and so files of commands can often be read by a pre-processor and then executed in sequence so that repetitive sets of instructions do not have to be executed by hand each time. Similarly, some systems allow these files of commands to act like computer programs, prompting the user for values and then having suitable logic to act on the input.

4.2.2 Solvers

Although called a solver, this program usually both sets up the required numerical equations that describe the behaviour of a structure under a given set of boundary conditions and solves the equations as well. The solver reads all the relevant data that has been defined by the pre-processor, usually held in files written by the pre-processor, then carries out the necessary numerical operations and writes the results to further files.

If necessary, the files can be moved between computers after being written by the pre-processor and after being written by the solver. This is extremely useful as it allows the solver program to be run on a machine designed to carry out numerical work, 'number crunching', at the fastest possible speeds and the pre- and post-processing programs to be run on machines designed for interactive use. This division of labour between machine types enables the hardware to be used more efficiently.

A further function of the solver can be data-checking. The solver checks to see if the data that it has read is acceptable before attempting to produce a solution. Examples of these checks include ensuring that the coordinates of all the nodes used to define elements exist and that the nodes defining an element are correctly oriented with respect to each other. Clearly such checks might also be carried out when the pre-processor writes the data for the solver.

4.2.3 Post-processors

As has been stated previously, these programs are often combined with pre-processor programs and have a common user interface. As large amounts of information are generated by the solver, graphical interpretation is often the only means of assessing the results. Hence the post-processor is devoted to the display of the results, with typical pictures containing, say, a section of the computational mesh together with vector plots of the stresses, contour plots of scalar variables such as stress intensity, or plots of the displaced shape of a structure when subject to a load. Again, extensive use of colour is made in these programs.

4.2.4 Utility Programs

Often utility programs are provided that are not part of the three programs discussed above. These utility programs usually convert the files of data written by one computer-aided engineering (CAE) system into a format that can be read by another CAE system. For example, information about the nodal coordinates and the element connectivity is sometimes created by one software package and used by another package that has a different data definition, or results data is written by a solver and then displayed in a post-processor unrelated to the solver. As integrated CAE packages become more widespread so the use of utility programs is reduced. Some people believe they are unnecessary, but at present this is not the case.

Sometimes files that are transferred between systems are referred to as *neutral files*, as they are standard text files that can be written and edited by system editors of many operating systems or by small programs that can be developed locally.

4.3 AVAILABLE COMPUTER HARDWARE SYSTEMS

4.3.1 Computer Architectures

From Secs 4.1 and 4.2 it can be seen that computer hardware is needed to act as a platform on which the software can be run. In the case of the solver program computational speed is important, but for the pre- and post-processing activities large amounts of data must be handled efficiently and computer graphics produced.

This appears to suggest that different computers are needed for the two generic tasks, but in many instances this is not the case as computers are now sufficiently fast that the solution of large engineering problems can be carried out on what might be considered small machines. Even though hardware changes occur regularly, a stable classification of hardware architectures can be used. Essentially, by considering features of the machines, the following five categories can be established:

Personal computers Such computers were developed as standalone systems with their own central processor unit (CPU), some random access memory (RAM) and disk storage. They also have a single-user operating system either resident in read-only memory (ROM) or on disk. More recently these machines have been networked together, with multi-tasking being possible.

Workstations These computers have a CPU, local RAM, true multi-user operating systems and are supplied with a high-resolution graphical display and, often, a reasonable amount of disk storage. Usually they are networked together so that they can access a central data storage system which might be a disk system attached to a *file server* which provides files to the machines on the network. Use of sophisticated networking software and special versions of solvers can enable the machines on the network to act in parallel—a form of distributed processing. Through the network access to high speed computers and peripherals is possible.

Minicomputers   These might be considered to be the traditional computer, comprising a CPU, large amounts of RAM and a central disk system. Such machines have multi-user operating systems and are accessed through terminals which have little of their own power.

Minisupercomputers   These are workstations with specially accelerated graphics performance and can also provide a good number-crunching capability.

Supercomputers   These are designed specifically to be efficient in running large numerical simulations. They consist of very high-speed processors, sometimes in multiple units, connected to a vast RAM storage system with fast access to reduce the time spent communicating with slow-speed disk storage.

Novel architectures   This category includes several types of machines, but two are predominant. Both use an array of processors, with one having a small number of expensive and fast processors and the other having a large number of inexpensive but slower processors. They are known as parallel machines, with the latter type having the special designation of massively parallel machine. Here a trade-off is made between computing power and cost, with the same total amount of processing power provided on both types for much the same overall cost. Unfortunately, such machines are at the forefront of technological development and so very little commercial stress analysis software will run on them.

Figure 4.3 shows the main architectures that have been described, but as technology changes, so the boundaries become increasingly blurred.

4.3.2   Peripherals

To achieve a good operational turnaround during an analysis, the user must rely on the total system, not just the raw computing power available. Peripheral devices such as the following can help speed up or slow down the process.

Secondary data storage devices   All engineering analysis programs that involve the numerical solution of equations such as the stiffness equation (1.2) throughout a multi-dimensional domain are required to access and store large amounts of data. Where possible this data is held in RAM storage during program execution as the response time is then the fastest available. However, secondary storage devices such as hard disks are also used both during execution when the available RAM is not large enough, to prevent repetitive calculations, and to enable data to be available to post-processors. As access to secondary storage is slower than with RAM its use can reduce the efficiency of any of the program types quite dramatically.

Backup devices   These are used to protect the data from being lost. This can occur when hardware, such as a disk drive, fails or when accidents, such as a machine-room fire, happen. Users, or more likely, system administrators, must

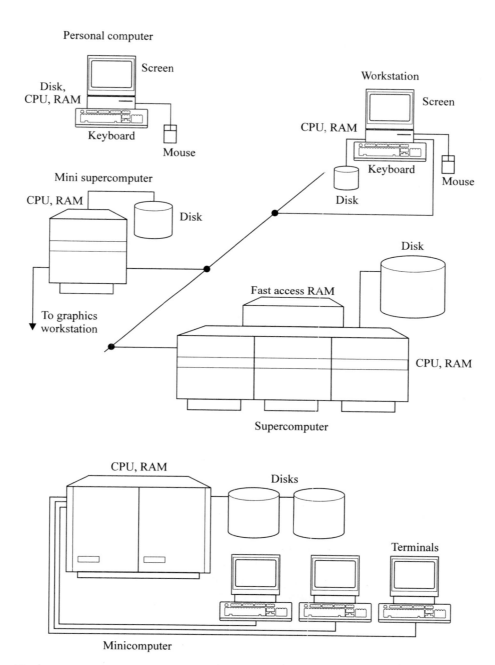

Figure 4.3 Types of computer hardware.

make regular copies of the data onto some form of backup data storage device. Typically, demountable hard disks or magnetic tapes are used as these can be removed to a safe storage area somewhere remote from normal operations.

High-resolution graphics displays As has been stated, graphical analysis using computer-generated displays in colour is vital to the success of large stress analyses. For user comfort high-resolution graphics display devices are used as the interactive display medium. Typically, the resolution of these devices is of the order of 1000 *pixels* in both directions, where a pixel is one dot of an array on the screen.

Hardcopy devices While looking at a screen is important to the analyst, hardcopy is also required so that sensible communication with others can be carried out. Such hardcopy is usually created using either laser printers which produce monochrome or colour paper copies or plotters which use ink jets or heated waxes to produce a coloured image.

4.4 MATCHING HARDWARE TYPES TO ANALYSES

Now an attempt can be made to match hardware and software for a variety of types of finite element analysis. First of all, computers can be classified by some measure of their calculation speed. This can be measured in units based on the number of internal processor instructions executed per second or the number of floating-point operations achieved per second. Common units are *mips* (millions of instructions per second) and *MFLOPS* (millions of floating-point operations per second). While these measures give some idea of the speed of a machine, they do not consider the complex interactions between a computer and the program being used. Such interaction is very important as the speed of reading from or writing data to RAM or disk can be just as important as the speed at which the numerical calculations are carried out. Further, each individual program is written such that the number of calculations is different for the same overall calculation and the amount of data access also varies. Hence it is very difficult to draw up hard and fast rules as to the turnaround time for an analysis.

Other considerations include the ways in which graphics data can be handled, in terms of both the speed of producing images and their resolution, as well as the amounts of data that can be stored. By considering these requirements—data storage, processing power and graphics capabilities—each of the computer architectures can be considered in turn to see to which types of analysis they might be best suited. For example:

- Personal computers (PCs) often lack fast processor speed or data storage capacity. Hence, they tend to be used for smaller problems that are simple to compute, such as linear statics analysis. An ideal situation might be in training analysts, where small demonstration problems are run and so the turnaround time is negligible.

- Workstations (or high-specification PCs) have the computing power, data storage and graphics capabilities for most finite element stress analysis problems where linear statics analysis alone is carried out. Equally, when optimization of a structure is required or dynamic or nonlinear behaviour is to be modelled, many problems can also be solved. It is only when very large and complex analyses are required that faster machines tend to be used.
- Minicomputers perform in a way similar to workstations although their graphics capability is often poorer.
- Minisupercomputers are versatile in that they can be used for most problems, even large and complex simulations. This is especially true if the turnaround time is not too important.
- Supercomputers are mainly used for the largest and most complex problems.

Table 4.1 shows a summary of these points. Here the graphics capability of the hardware types is rated together with the applicability of using these types of hardware for small, medium and large analysis jobs. Clearly, the definitions of the size of a job are somewhat arbitrary, but small jobs could be those with only a few hundred to a few thousand nodes for linear statics analysis of say a two-dimensional or axisymmetric situation. Medium jobs could be those with a few tens of thousands of nodes for linear statics analysis and large jobs might be those with tens of thousands of nodes or more and those where more complex analyses such as time dependent or nonlinear simulations are being carried out. Note that some structural analysis can, in most cases, be carried out with only a limited amount of hardware.

4.5 ACQUIRING THE TECHNOLOGIES

4.5.1 Some Preliminaries

By using the hardware and software that is commercially available, designers have access to a tool that is complementary to physical experiments, semi-empirical methods and analytical techniques in analysing proposed or existing structures and might lead to cost reductions. However, the acquisition of the necessary hardware and software, as well as the appropriate human expertise, is expensive.

Table 4.1   Suitability of hardware types

Hardware type	Graphics	Small jobs	Medium jobs	Large jobs
PC	OK	OK	No	No
Workstation (or high-spec. PC)	Good	Good	Good	No
Minicomputer	OK	Good	Good	No
Minisupercomputer	Good	Good	Good	OK
Supercomputer	No	OK (expensive!)	Good	Good

For this reason, a careful study of the needs and requirements has to be made before the final decision to acquire the technology can be taken.

4.5.2 Choosing Software

The first thing to consider is how a knowledge of structural mechanics might help you or your organization. To explore this, all the functional areas that are related to structural mechanics must be considered. Take the case of a manufacturer in the motor industry. Structural analysis may be used to determine the linear static stress and displacements in structures such as the vehicle body shell and the engine under operational loads. Also, optimization may be required to produce body shells with a given displacement for the minimum material thicknesses. Further, coupled heat transfer and structural analysis may be used in analysing the distortion of the braking systems and the engine block and, finally, crash simulation of the body shell may be required.

Once the areas of interest have been listed, the techniques that are available to investigate the particular type of analysis should be assessed. If computational tools have a place in the toolkit of the designer then the benefits of using the technology must be made clear. Remember that computational tools may not provide more information than experimental procedures, but they may provide some extra benefit, such as a direct saving in financial terms through the reduction in the number of physical models required, or the ability to carry out more tests in a given time.

If there is a clear need for structural analysis software, then the type of software that is required must be decided upon. To find out which of the available packages may be used, a list of requirements that the software should meet must be produced. More often than not, no single package will meet all the requirements, but several packages will meet some of the requirements. Hence, when choosing the software analysts may have to make some very subjective decisions.

To draw up the program specification, analysts must think carefully about the structural problems that they wish to analyse and the priority that is to be attached to them. For example, consideration might be given to the following features:

- *The geometry of the structures that may need to be analysed* This will show whether a package is needed that can solve problems in two or three dimensions. Invariably, three-dimensional solutions will be required, but the analyst needs to see if the geometry is very complex, when an unstructured mesh is required.
- *The structural analysis type* Here the analyst must decide which type of analysis needs to be carried out. This could be simple linear static analysis or it might include a requirement for optimization. Equally, the simulation may need to be time dependent or have to calculate the effects of nonlinearities such as plasticity and/or buckling.
- *Coupling requirements to other software* In some cases there may be a need to link structural results to heat transfer simulations or even to fluid flow soft-

ware. There are also other interfacing requirements. For example, the software may need to have access to geometrical data created or stored by a proprietary CAD system. Equally, there may be a requirement to send the results to a proprietary post-processor or to some other display software. Clearly, in these cases, the structural software must have appropriate interfaces.

- *The size of the simulation problem*   Here something about the number of nodes and elements that a typical mesh contains needs to be known, together with the number of degrees of freedom that are to be calculated. This information helps to determine the storage requirements of the programs in terms of both primary and secondary storage, as discussed in Sec. 4.5.3. Remember that the more data that is calculated, the more accurate the solution should be, but the longer it will take to obtain the results. Clearly some compromise has to be made here.
- *The required results of the analysis*   Stresses, strains and displacements, possibly as a function of time.
- *Solution speed*   Many things affect the time that it takes to produce the solution. Clearly, this depends on the processing speed of the hardware that is used, but it also depends on the structural solver itself. Some algorithms for solving the governing equations are much faster than others. This speed difference can come from the basic discretization of the equations, from the internal organization of the program or from the speed of the linear equation solvers used.
- *Hardware availability*   If there is a restriction on the make or type of computer or graphics terminal that the software can be run on, this should be noted. It is common, however, for pre- and post-processors to be very hardware specific and for solvers to be much more portable between different machine types.

From all of the above, a considerable amount will be known about the situations that are to be simulated and the simulation process itself. Finally, it is important to assess what kind of service it is that the software supplier must provide. This is a very subjective set of requirements, and to some extent depends on the people who are available within an organization to run the analysis software and talk with the supplier. At this point, it is worth issuing a warning. With the proliferation of computers and software, many people are now used to buying a package, loading it onto a machine and getting results without too many problems. For business software this is certainly true, and it is becoming true of many engineering packages as well. Unfortunately, any software that solves partial differential equations will not, by its very nature, be as mature as other simpler engineering analysis tools that are on the market. Hence, in most cases, the learning curve will be very steep and costly.

The following are some of the requirements that are related to the software supplier:

- *Quality assurance (QA)*   This is the extent to which the software has been tested against standard test cases for which there is a known solution. The comparison data may come from either analytical expressions or experiments.

These comparisons are carried out to both verify the code, i.e. to show that it is correct against the numerical models that it is simulating, and also to validate the code, i.e. to show that the code gives reliable results against physical experiments. There should be some evidence from the supplier that the software has been tested for both validation and verification. In fact, most pieces of structural analysis software are so complex that every possible combination of operating features will never be tested until, that is, you as a user run your particular example and find that it does not work. This may appear cynical but it is often true.

- *User friendliness* This is probably the most subjective feature of all, as what appears friendly to one person may be unfriendly to another. Again, this depends on the staff who run the programs. However, it is worth looking at the user interface to see if the menu structure is logical.
- *User support* This is very important as users can never be fully conversant with the programs that they run. Software suppliers should provide some form of user hotline that can give a quick response to a user's questions. This normally comes as part of the annual licence fee for the software, or can be purchased separately if the program is bought with a once-only payment, which is known as a perpetual licence. There should also be the option of buying training in the use of the programs, and the chance for users to work with the supplier in setting up a problem. This is normally done by paying for consultancy from the software supplier.
- *Current users* It is important to know who *is currently using* the software, not who *has used* the software. This enables companies to see if firms in a similar business are using the software, and can give some confidence in the software supplier and their product.

The easiest way to document this information is to draw up a table of capabilities product by product. Sometimes this can be difficult to do for someone with little experience. It is at this stage that it is important to obtain independent advice to guide you. Sometimes people place too much reliance in the software suppliers themselves and, even though the suppliers can provide much information, an independent view is worth while. A sample specification table is shown in Table 4.2 and this can be used to assess each of the competing products.

Once the specification of the software is determined, various software options need to be evaluated against this specification. This can take a lot of time and effort as there are many products in the market-place and the suppliers of each of them will be only too willing to shower you with information. The information that is provided can take many forms, but the simplest starting point is to look at the brochures that explain the software. Much of the information required can be determined from these, but quite a lot of it cannot. In particular, the more subjective information such as the levels of user friendliness, solution times, QA and user support need to be investigated further.

One way of gaining this more specific information is to produce a sample problem that is typical of those you wish to solve. Suppliers will often produce a

Table 4.2 Assessing a program specification

Capability	Prog. 1	Prog. 2	Prog. 3	...	Prog. N
Dimensions (2 or 3)					
Mesh type (structured or unstructured)					
Linear statics (yes or no)					
Normal mode dynamics (yes or no)					
Buckling (bifurcation) (yes or no)					
Forced dynamics (yes or no)					
Large deformation (yes or no)					
Plasticity (yes or no)					
Optimization (yes or no)					
Time dependent (yes or no)					
Heat transfer (yes or no)					
Coupling to fluid flow (yes or no)					
Limits on model size					
Interfacing to CAD					
Speed of solution					
Hardware availability					
Evidence of QA					
User friendliness					
User support					
List of users with similar structural problems					

simulation of this problem using their software at a reduced cost, or even for free if the problem is very small. This enables potential customers to see software products in action on a realistic problem. Such a trial helps in understanding how the processes outlined in this book relate to the specification and operation of the software. It also produces some hard facts that should help in determining the cost of obtaining a simulation using a particular package.

When the competing products have been assessed using the specification table, several suitable products should emerge. It is probable that none of the products will be ideal, but some should come closer than others. A simple way of assessing the most suitable package is to assign numbers to each of the categories in the specification in some way such that the higher the number the better the specification level. By adding up these numbers and getting a total value for each package they can be ranked.

Once the products are ranked in order of suitability, the question of cost needs to be looked at. Normally, software is licensed on an annual basis with a single fee being paid to the supplier which includes the provision of the software and any updates to it as well as technical support in the form of a hotline service. Sometimes, however, the software is purchased on perpetual licence terms where one large payment pays for the software and a smaller annual fee pays for the updates to the software and the support. Sometimes both methods are on offer, and it takes careful consideration to decide which of the two will be the cheapest option in the long run. This is especially difficult as the market is still developing and the most suitable program today may not be the best choice in three or four years. Finally, it may be that some sacrifice in terms of the capability of a package has to be made if an affordable solution is to be chosen for purchase. This is achieved by determining the minimum level of functionality that is acceptable.

4.5.3 Choosing Hardware

Many organizations already have access to the computer facilities that are necessary for running large computational analysis programs. Others will need to acquire the hardware. In both cases, however, it is important to consider the following factors. For the former case these factors enable the user to determine if the existing facilities are suitable and have the necessary spare capacity, and for the latter case they allow estimates to be made of the various measures that will determine the hardware.

Computer processing power   A large amount of processing power is needed to run some structural problems. Fortunately, recent technical advances mean that the necessary power is available cheaply. The factors that affect the speed of processing include such things as the calculation speed of the processor which is measured in mips or MFLOPS. There is no clear relationship between the two for different processors, as what takes one instruction on one processor might take several instructions on another. The speeds of the various computers are often quoted in these units, but different software runs in different ways on different machines. Consequently, the numbers quoted are only a guide to the raw processing power. To find a true measure of speed for the software and hardware combination a series of sample problems must be simulated. This assumes that the software does not make any use of the secondary data storage during execution, as the speed at which data can be accessed from devices such as hard disks can have a marked effect on solution times. Some software packages write data to these devices during the solution phase and if the processes of reading from and writing to the disk are slow, then the whole solution process is slowed down. For a typical analysis on a given computer installation, the total solution time depends on all of these things together with the number of simulations that are solved simultaneously on any one system.

Primary data storage capacity   On most computer systems the primary data storage system is RAM. This is usually sized by the number of bytes of data that can be stored. Each byte consists of eight bits, where one bit is the basic unit of storage corresponding to a stored value of either zero or one. Numbers can be stored as integers or real numbers and two or four bytes are used for integers and four or eight bytes for real numbers. The greater the number of bytes the greater the maximum integer, and the more accurate a real number, that can be stored. Sometimes, the software supplier specifies the number of megabytes of RAM that are required to run their software successfully. In large machines, such as supercomputers, the memory size is measured in words. These are usually words of eight bytes or sixty-four bits, and are the machine's minimum storage for a single real number.

Secondary storage capacity This is usually provided by hard disks, which are aluminium disks covered in magnetic material such as iron oxide, just as with audio tape. In personal computers these disks may store tens of megabytes and in workstations several hundred megabytes. In large systems, the disk storage might consist of sets of disks each storing several gigabytes of data. Analysts need to assess how much of this storage is needed for each problem to be solved. A rough estimate can be made by taking the number of nodes in a problem and multiplying this by the number of coordinates used to describe the spatial position of a node plus the number of variables stored at each node or element. For a three-dimensional linear statics problem there are 3 coordinate directions and 3 displacements per node giving 6 variables at each node. So for a mesh with 10 000 nodes, at least 60 000 real numbers must be stored. If the data is stored in readable (ASCII) format, say 20 bytes (or characters) are required to store each number, and therefore 1.2 megabytes are required in total. If, however, the data is stored as single precision real numbers in binary format, only 4 bytes are required to store each number and the total storage required is 0.24 megabytes. These are low estimates of the total data storage requirements, as each software package will store different information, but a software supplier may be able to give information on the data storage required for a given model size.

Access points If several people need to run analyses simultaneously then several access points are required. These may need to be split between a number of graphics screens and a number of text screens. This enables some people to perform graphics pre- and post-processing, while others run the solver program.

Backup facilities There is a need to provide some backup of the data held on disk, to protect against loss of data. This can occur if a disk drive is broken, such that the data stored on it cannot be read, or can occur if a user deletes a file in error. It is common for each disk to be backed-up in full, i.e. all the data is written to a tape storage device, or something similar, every week. Further backup procedures are carried out once a day, to ensure that all the new files that are created within the previous 24 hours, and the new versions of edited files, are also written to a backup device. This procedure is known as an incremental backup and ensures that, at worst, only one day's work can be lost. Once backup tapes have been prepared it is worth protecting them against fire by using a fireproof storage facility.

When these items have been considered, it should be possible to know whether an existing installation will be sufficient to run the relevant structural problems or whether it will need to be enhanced in some way. If new facilities are required, either to enhance the existing capacity or to provide a completely new system, then they can now be assessed for suitability.

4.5.4 Building an Analysis Team

Having decided upon a software package and a hardware system, it must be made clear that the simulations do not run themselves. Consequently, some consideration must finally be given to the most important asset in the analysis process. This is the analyst who actually translates the engineering problem into a computational simulation, runs the solver and both checks and analyses the results. It is the skill of this person, or set of people, that will determine whether all the hardware and software is utilized in the best possible way and so produce good quality results.

To produce a successful and cost-effective linear static finite element model the analyst has to be capable of the following:

1. Understanding the fundamental physics of the problem so as to gain a good impression of the expected solution. Here physical interactions between the structure and its environment are the principal difficulty. A working knowledge of classical elasticity, as presented in Chapter 2, is most useful as it allows the analyst to determine where in the structure the stresses, and their gradients, will be highest. Load paths through the structure can also be visualized. Also the analyst needs the ability to identify regions of stress concentration and determine die-away lengths (using Saint-Venant's principle) associated with a given type of discontinuity.
2. Having the skill to choose the relevant dimensionality and the structural form for the preparation of a satisfactory model. This also requires an understanding of the limitations of the applicable theory and so, again, there is justification for knowing the theory of classical elasticity.
3. Understanding the relevant facilities of finite element packages and how these are used to model the various features of the physical problem.
4. Understanding the behaviour, accuracy and limitations, such as 'locking' and mesh instabilities, of the various finite elements available, the approximations made in representing real boundary conditions (loading and restraints) and the solution algorithms.
5. Having the skill to construct a mesh with the minimum number of degrees of freedom that provide the desired accuracy while preventing round-off errors from becoming intolerable. This includes the application of techniques such as assuming symmetry, assigning multi-point constraints and substructuring.
6. Having the skill to choose relevant and reliable material property data and, if required, failure criteria for the material used in the structural problem.
7. Having the skill to develop a reliable and relevant model that meets the goals of the analysis.

8. Modelling the real boundary conditions of loadings (mechanical loading, thermal loading, displacement conditions or initial conditions), supports, constraints, joints, gaps and releases, and ensuring that all rigid body motion is removed without altering the structural response.
9. Checking, before the solver is run, that the model is valid.
10. Interpreting output results sensibly and without reliance on powerful graphic post-processors that can smooth out stress discontinuities that may be warning the analyst that the model is wrong.
11. Comparing solutions with whatever else is available, such as 'back-of-envelope' calculations, approximate analytical models, experimental data, textbook and other computationally generated results, which may well be based on a different numerical method.
12. Refining the model using h- , p- or adaptive mesh refinement with the minimum of work to determine the rate of convergence to the 'exact' solution.
13. Using the results from the analyses to make the 'correct' design decisions, what ever they may be.

This comprehensive list of skills and knowledge that an analyst requires to be a successful user of static finite element analysis in the design process illustrates the many difficulties that have to be faced. Many of these will be addressed in later chapters. Note that if the structural problem requires analysis of dynamics, buckling and/or nonlinearity, then the list of skills and knowledge will be increased significantly from that given above.

People from many different backgrounds can be trained to carry out the processes listed, but it is our belief that more than this is required. The successful application of the finite element method (FEM) requires not just basic knowledge but a considerable degree of experience on the part of the analyst which can only be obtained by practical use of the finite element model.

Another way of looking at the skills required is to consider broad categories of skills across four areas.

Mathematical skills   These enable the analyst to understand the underlying features of the numerical processes used to convert the governing partial differential equations into numerical analogues, and to coax the solution procedures to converge to sensible and realistic values.

Computational skills   The production of a simulation can involve the user in manipulating large amounts of data with packages that do not interface together and reside on a variety of types of computers. This can mean, for example, that analysts have to write their own interface programs to convert data from one program's format to another program's format. Also, an analyst might have to write computer operating system command language programs that instruct a computer or even a variety of computers to move data around a network, run some programs and then move the data around the network again. Consequently, analysts must be conversant with computer procedures at a level that is far greater than that required for analysts who use the more common software products that perform engineering computations.

Good interpersonal skills If the analyst is not the end user of the data, then there has to be close liaison between the analyst and the end user, who is in effect a customer or client of the analyst. This requires that a good working relationship is developed between the two parties so that the analyst knows what the customer requires, and the customer is aware of the limitations of the analysis.

Engineering skills Finally, the analyst must have a working understanding of the engineering processes that are to be modelled. This enables the limits of a computer model to be established and the results of the simulation to be analysed in a sensible way.

Large organizations may well have a pool of analysts in which there are several people who could be used to produce structural simulations, as they have a majority of the qualities listed.

If such people do not exist within the organization, or if suitable people cannot be used for whatever reason, then staff will have to be hired. Hiring staff of the right technical background to use finite elements in industry, whatever their background, is extremely difficult. Not many people have all the skills necessary and so several people may be needed. Depending on the size of the organization, therefore, one or more people may be employed in the use of finite elements, and the right mixture of abilities is important.

One other way of proceeding is to employ a limited number of people to work with finite elements and then to use external consultants to supplement the skills where appropriate. These consultants can be found working with software suppliers, general engineering consultancy practices and in universities. For industrial users who are not specialists in this field, it is important to have access to advice at a moment's notice. This can be provided by a software supplier when problems occur running a particular package, but a local university may well have a specialist in the finite element field who may be willing to provide consultancy as and when required.

5

BUILDING THE GEOMETRY DESCRIPTION AND THE MESH

This chapter contains material previously published in Shaw (1992), by kind permission of Prentice Hall.

5.1 THINK FIRST—COMPUTE LATER

When using a computer to simulate any engineering problem an analyst must specify the model in the form of data to the solver. This data must reflect the actual engineering problem being solved. While the data must be both suitable and accurate, it may be that other factors such as a user's experience will determine the quality of the simulation. Given that this is the case, an analyst must become familiar with the engineering problem that is to be simulated. While it is a common failing for those who use computers to be tempted to start computing as soon as possible, it is much better if careful thought and planning is given to the problem first. Hence the urge to compute before thinking must be resisted.

Particular problems arise when people use computers as black boxes. The well-known computer acronym GIGO applies here—garbage in, garbage out—so users of finite element packages for mechanical design should be thoroughly familiar with the material in Chapter 2 before they embark on a static analysis. In particular, the thinking required to develop a good feel for a problem is given in Sec. 2.4, where a problem specification is produced that is the main source of information for the analyst. It is the information contained in the specification that has to be translated into terms that the software package can understand.

Once the specification of the structural problem is known, attention can be turned to the next stage of the analysis process, where the mesh is generated. When modelling a simple problem this process takes very little time but, when modelling a complex geometrical problem such as the body shell of a motor vehicle or a casting for a piston engine, the process can take one engineer several months to complete. Often it is this stage of the analysis process that determines the total time required to obtain results from a simulation, as all the other phases, including the actual computation of the results, can be carried out quite quickly. Similarly, the overall cost of the analysis can be totally dominated by the costs of the labour required to build the mesh.

In fact, the mesh generation stage can be subdivided into the building of a computer model of the geometry that is suitable for use in the mesh generation process and the creation of a mesh to represent this geometry. Both subdivisions will be covered in this chapter. First, the ways in which geometry can be modelled using computers will be discussed before looking at the engineering approximations that can be made. Then the reasons for building a mesh, the choice of appropriate element types, the requirements that a mesh must satisfy if it is to give satisfactory solutions and the types of mesh that can be built will be discussed. Finally, the ways in which a mesh can be built using a variety of software tools and the ways in which a mesh can be modified in the light of the results of a simulation such that better results are achieved will be considered.

Throughout this chapter it should be kept in mind that it requires considerable skill and a good deal of experience to be able to construct a mesh that will produce an accurate solution for an acceptable cost for even a moderately complicated structure.

5.2 OVERVIEW OF COMPUTER MODELS OF GEOMETRY

In the specification stage of the process that was discussed in Sec. 2.4 it was seen that the sources of geometrical data must be determined and sketches produced of the structure. For structures with essentially one-dimensional elements such as bars and beams this includes the positions of these elements, but for more complex structures this includes the positions of the bounding surfaces. These sketches must be used together with the sources of geometrical information to produce a computer model of the geometry.

While it is not necessary to understand the mathematics behind the descriptions, the analyst should have some knowledge of the variety of computer models that can be used. In particular the following three levels of model are often used:

- wireframe models
- surface models
- solid models.

The level of complexity increases when going down the list, with the first handling just curves, the next handling surfaces and the final one handling the internal volumes as well. For those who are interested, several books describe the ways in which these numerical descriptions of objects are handled in CAD systems, for example, Rooney and Steadman (1987) and McMahon and Browne (1993).

5.3 WIREFRAME MODELS

A special case that needs to be dealt with here is the class of structures that can be thought of as consisting of one-dimensional elements. These are special as the geometry is completely described by the end coordinates of the bars and beams, together with the connections between them. Hence while computer models are of the wireframe type, they are different in nature to the classical wireframe models.

For two-dimensional problems, the bounding surfaces can be created using points to define a series of lines and curves. These curves might be defined as circular arcs, simple polynomials or splines. All of these constructions are described by equations that define the relationship between the coordinates of points that make up the curve. For example, it is known that a line can be described by the relationship

$$y = mx + c \tag{5.1}$$

where m is the gradient of the line and c is the value of y when x is zero. By substituting for the gradient and intercept in terms of two known points on the line, (5.1) becomes

$$y = \frac{y_2 - y_1}{x_2 - x_1} x + \frac{y_1 x_2 - x_1 y_2}{x_2 - x_1} \tag{5.2}$$

where the subscripts refer to the two known points. These equations describe a line that is infinite in length but, as lines of finite length only are used to describe the geometry of a plane structure, the endpoints of the line must be known.

Similarly, a circle can be described by

$$(x - a)^2 + (y - b)^2 = r^2 \tag{5.3}$$

where the centre of the circle is at $x = a$, $y = b$ and the radius is r. A part of this circle is a circular arc and three points can be used to define it. Usually these points are taken to be the two end points of the arc and a point on the arc somewhere between them. This enables both the limits of the arc to be defined and the unknown constants in (5.3), namely a, b and r, to be calculated.

Splines are more complex curves but they are also defined by points in space. Usually four or more points are used, and these do not have to be on the curve itself. Note that there is a hierarchy being formed here in terms of the numbers of points required to define a curve. Two points define a line, three points define an arc and four or more points define a spline.

Clearly, in three-dimensional problems, a similar hierarchy exists and so points in x, y and z can be used to define three-dimensional lines, arcs and curves.

These in turn can be used in the form of a *wireframe model*. In such models key edges of the model are defined by points in space, i.e. the model consists of a series of edge–vertex relationships. Here, several lines, arcs and curves can make up a single edge.

Typical uses of this type of geometry model in mesh generation are for simple meshes in two dimensions or for simple three-dimensional meshes such as a sheet of nodes and elements through space, as might be used to model a topologically rectangular region of the surface of a structure.

5.4 SURFACE MODELS

In many cases a wireframe model is not sufficient for mesh generation as detailed information is required as to the bounding surfaces of the geometry. While it is possible for a structure to have surfaces that are of simple form, such as a plane or part of a sphere or cylinder, the surfaces of a structure often have a more complex form. When this is the case the complex surface is often described by a series of simpler surfaces which are local approximations to the physical surface.

Typical approximations to complex surfaces are simple patches, Coons patches, Bézier surfaces and non-uniform rational B-spline surfaces.

Numerous simple patches   Here the surface is discretized into a series of patches that are usually triangular or quadrilateral in form. This is the way that a surface is described by the faces of a mesh of linear elements.

Coons patches   These are patches over which the coordinates of points on the surface are determined from the bounding curves alone. Consequently, once the boundaries of a surface are determined the surface itself is defined. Three or four curves in space, or assemblies of curves (i.e. edges), which form a closed loop are often used to define the boundaries. Note that an infinite number of surfaces can fit through a given set of boundaries but that the Coons patch description defines only one such surface. The assumption has to be made that each patch represents a sufficiently small part of the surface so that a good approximation to the surface is given. This approach can lead to problems if a surface is highly curved and only a few Coons patches are used to model it. In this situation each patch will be too large and the surface definition will depart dramatically from that required.

Bézier surfaces   These are surfaces described by a set of Bézier polynomial curves. Each curve is defined by four points, the two end points of the curve plus two interior points which need not be on the curve. By moving the two interior points the curve can be manipulated to have a wide range of shapes. To construct a Bézier surface, 16 points are used to define a 4 by 4 lattice of Bézier curves and by interpolation between these curves the surface coordinates can be found. Bézier surfaces give an improved description of a surface when compared to a Coons patch description, as information from within the boundaries is used to define the surface. This helps to lock a surface in space and so the number of surfaces that can fit the description is reduced. These surfaces were developed for Renault, the French vehicle manufacturer, as they required computational surfaces that could be manipulated interactively when modelling new vehicles in the styling studio.

Non-uniform rational B-spline surfaces (NURBS)   These are similar to Bézier surfaces, but the curves that are used to define them are based on different points to the Bézier curves. The end points of the curves are only approximated but the points that are used to define the polynomials ensure that the spatial derivatives of the first- and second-order are continuous at the end points.

5.5   SOLID MODELS

Many CAD systems now incorporate both wireframe and surface models together with the third model type, the *solid model*. Here, not only geometric information but also such data as the mass of an object is stored. To produce this information two main types of methodology are used, *constructive solid geometry (CSG)* and *boundary representation (B-rep)*.

5.5.1   CSG

In CSG complex objects are built from combinations of primitive objects such as cylinders, tubes, blocks and cones. By performing the Boolean operations of union, intersection and difference, a graph or tree data structure can be developed. For example, Fig. 5.1(a) shows a thick plate with a cylindrical hole through it. This object can be considered to be built up from the union of two plates and the difference between the union of these plates and a cylinder as shown in Fig. 5.1(b). This data structure is known as a CSG tree.

5.5.2   B-rep

In B-rep models the objects are defined by their boundary descriptions. For example, a closed loop of curves, also known as a profile, might be extruded through space to form a three-dimensional object, as shown in Fig. 5.2(a), where an extrusion along the z-axis is carried out for a profile defined in the x-y plane. Another common operation is to sweep a profile through some angle to form a volume of revolution, as shown in Fig. 5.2(b).

More complex B-rep models can be created by taking a series of profiles that have been defined along some path in space and then effectively placing a surface over these profiles. This operation is known as skinning and an example is shown in Fig. 5.2(c) where three circular profiles are placed with their centres along some path in space and a skin thrown between them. The process works in a similar

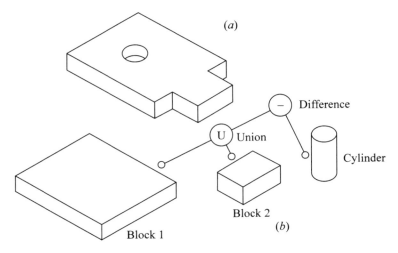

Figure 5.1 A CSG model of a plate with a hole: (*a*) the final object, (*b*) the CSG tree of Boolean operations.

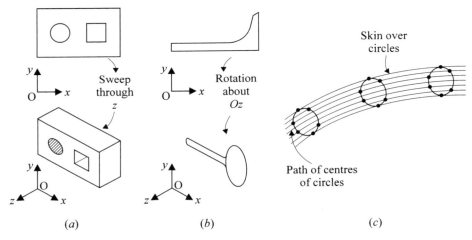

Figure 5.2 Operations for B-rep models: (*a*) sweeping a two-dimensional profile in the third dimension, (*b*) forming a solid of revolution, (*c*) skinning across a set of profiles.

way to the lofting of ship hulls where complex curves are placed along the length of the ship to define the hull shape.

5.5.3 Commercial Solid Modellers

Most commercial solid modellers combine the features of CSG and B-rep to give a more flexible modeller. Such systems allow CSG-type operations to be carried out on a range of solids which can be both primitives and B-rep models. Occasionally, other types of solid model can also be used such as pure primitive instancing and cell decomposition (Rooney and Steadman, 1987).

5.6 CHOICE OF GEOMETRY DESCRIPTION

Several factors affect the way in which the analyst chooses to describe the geometry. These include the following:

- The *original sources* of the geometry data. For example, if there is no computer description of the geometry the analyst is free to use whichever description is appropriate, subject to the factors that follow. Equally, if the geometry data is part of a computer database already, the analyst will want to use that data in as efficient a way as possible, obtaining the appropriate information with the minimum amount of effort.
- The *tools available* to the analyst. Clearly, these might restrict the approaches that the analyst can take if software is not available.
- The *physical problem* being modelled. Where the use of dimension reduction to model a three-dimensional problem as a one- or two-dimensional problem or the use of symmetry can be very effective in saving computer effort.

It is the last of these three factors that can make or break an analysis. Remember that three-dimensional analyses are at least an order of magnitude more expensive than two-dimensional ones and that, because there are many more equations to solve, such simulations are prone to round-off errors. Further, it is found that three-dimensional models present problems in mesh generation, because of large element distortion, and in checking, because of poor element connection providing unwanted cracks that are difficult to detect and rectify. Additionally, while the subdivision of a two-dimensional domain by triangles is easy to visualize, the subdivision of a volume by tetrahedra is extremely difficult to visualize.

As a guide:

> Use three-dimensional models only if either one- or two-dimensional approximations have been found to be unacceptable.

Equally, the use of symmetry is advantageous in that it leads to a reduction in the effort required for model development, a reduction in the required computer power, computing time and cost and to a decrease in the computer round-off error when solving the equations since fewer equations exist in the model. However, there are disadvantages in the use of symmetry: it can be more difficult to picture the model and the peak stresses may occur along symmetry surfaces making it more difficult to locate the peaks accurately.

If the problem has symmetry of material properties and boundary conditions then one or a combination of the following four types of geometric symmetry can be exploited:

● *mirror symmetry,* where the geometry can be modelled by one half of the full geometry as there is a plane of symmetry
● *axial symmetry,* where the problem is such that it can be modelled as a quasi-one- or two-dimensional section in some radial coordinate and a physical length coordinate
● *cyclic symmetry,* where the geometry and boundary conditions allow one part of the geometry to be modelled as a representative section of the whole geometry
● *repetitive symmetry,* where sections of the geometry repeat themselves.

Of these, mirror and axial symmetry are fairly common, but cyclic and repetitive symmetries are not. Hence, for the last two types, a significant degree of understanding is required to apply them. Further details can be found in NAFEMS (1986).

5.7 OVERVIEW OF MESH GENERATION

5.7.1 The Need for a Mesh

In Chapter 3 the finite element method was considered as a way of discretizing a structural problem so that numerical equations describing the deformation of the structure are produced. To do this requires the discretization of the volume of the structure under consideration such that a set of points, known as the nodes, is created that defines the subregions of the structure, known as elements.

The simplest meshes are for those structures that can be considered to consist of one-dimensional members such as beams and bars. These structures can be easily described by nodes and elements generated from the locations of the end points of the members. In subsequent sections of this chapter only meshes for more complex problems which are described in two and three dimensions will be considered.

5.7.2 The Constituent Parts

From the discussions in Chapter 3, it is already known that a mesh consists of subregions known as elements on which the variables are found at fixed points known as nodes. These then are the basic parts from which any mesh is built. To define any mesh the nodal coordinates must be calculated in the appropriate coordinate system, and then the elements themselves can be defined by listing the nodes that are attached to each element. This list is known as the *connectivity list.* At this time it may be that the material properties of the element also have to be given as well as a flag to denote the type of element and its mathematical description. A variety of element types are presented in Table 3.2.

Knowing that nodes and elements need to be created, it must now be decided which elements are to be used.

5.8 ADVICE ON CHOOSING ELEMENT TYPES

5.8.1 The Range of Elements and Testing the Elements

The available elements can be listed in the following six categories:

- membrane elements (and plane stress and plane strain)
- bending elements
- general shell elements
- solid elements
- axisymmetric elements
- specialized elements like springs, gaps, rigid bars and elastic foundations.

It is not possible to present a set of universal guidelines to develop any finite element model as each structural problem and element type have their own particular features. It is not even possible to give rules for what appear in packages to be identical element types since their formulation can be different. As an example, some packages contain the six-noded triangular element described by (3.25), while others collapse a line of nodes of an eight-noded quadrilateral to form the triangle. Although the two elements are geometrically the same they behave differently when loaded.

To determine the behaviour, or performance, of elements, it is up to the analyst to perform a series of simple tests known as *patch tests*. These are implemented by seeing if a group of elements, which are usually distorted, can model constant stress and be strain-free when subjected to a rigid-body movement.

As a guide:

Any test for element behaviour should be more complicated than the situation of a simple rectangular geometry with a constant load, since simple situations can give a false impression of the convergence characteristics for realistic problems.

Such tests are particularly important in developing an understanding of the behaviour, accuracy and limitations (e.g. 'locking' and mesh instabilities) of the various finite elements available, as well as the approximations made to represent real boundary conditions (loading and restraints) and the solution algorithms. Here the performance of any element must be established against a known solution. This must always be performed for plate and shell elements because of the large number of formulations and the fact that the vendor's documentation is unlikely to describe fully the formulation and features of its available elements. It is difficult to make a good judgement about which element to use unless simple tests are conducted. By word of warning, most of the research effort has focused on element formulations and convergence criteria, but the test cases for evaluation are usually small-sized academic cases (including patch tests) rather than actual problems. So the demonstrations of adequacy in the research cases may not be very helpful in the practical cases where an analysis must be done. NAFEMS have developed a series of benchmarks which are a valuable aid.

Current practice indicates the following:

- Quadratic elements, be they membrane or solid elements, give the best compromise between accuracy and efficiency for general use.
- Even though a finite element package has the linear tetrahedron solid element (see Table 3.2) in its library it should not be used because it has been condemned by the analyst community.
- When modelling a structural problem that can be classified as having bending (with or without membrane) deformation and the geometry is either flat or curved, then the preferred choice of element is always the general shell element based on the isoparametric Mindlin theory.
- Curved surfaces should not be modelled using flat elements as the discontinuity at element boundaries introduces significant error.

5.8.2 Using a Hierarchy of Elements

When approaching a new type of problem the experienced analyst should perform a parametric study to investigate combinations of structural and element behaviour related to the problem. For instance such studies can be used to determine the effect of die-away lengths.

Another example is the use of an approximate two-dimensional model before building a three-dimensional model. This undoubtably saves resources as the two-dimensional model provides invaluable information for the three-dimensional mesh specification in terms of the mesh density.

As a guide:

Analysts should develop a model using a step-by-step approach. This means that they should start with a simple approximation, say a beam model, and make it more precise as the finite element modelling progresses. Never tackle a real problem directly as this is likely to be time consuming and wasteful of resources. Remember, the more results that are generated the more effort that will be necessary to check that they are reliable and relevant.

Equally, results from one model can be used to provide boundary information for another. An example of this is that the results from a general shell model may be used to feed information to a model generated with three-dimensional solid elements of a local region of interest. This is a process known as *substructure modelling*. In this case it is necessary to plan the shell mesh so that a local group of elements will provide a suitable set of boundary information for the local solid model.

5.8.3   Restricting the Dimensions of a Problem

By using plate or shell elements instead of solid elements not only is the computing effort reduced, but also the accuracy may be improved. As a guide:

> Avoid the use of solid elements to model a problem where the length in one of the spatial dimensions, for example the material thickness, is much less than the lengths in the other two dimensions.

This reduction in accuracy with solid elements is due to ill-conditioning. To achieve acceptable aspect ratios for the elements the mesh will require many elements and Eq. (1.2) will therefore have a large number of degrees of freedom. NAFEMS (1986) illustrates this feature using the example of a simple cylinder whose radius is 10 times the wall thickness. Here the choice of element can be either axisymmetric thin shell, axisymmetric thick shell, general thin shell or general solid.

As an aside, note that with computing power growing rapidly there may come a time in the not too distant future when the use of solid elements in this situation does become realistic.

5.8.4   Plate and Shell Elements

Plate and shell elements have historically been the most difficult to use in terms of achieving reliable and cost-effective solutions. At this time, there is no particular element that is broadly acceptable within the analysis community.

In particular these elements in a static analysis do not give an acceptable solution if the displacement of the nodes normal to the surface of the material is greater than the thickness of the material.

The following guidelines may also be of use:

- If the structural problem consists of flat panels of material connected into open or closed sections (e.g. I-sections, box sections or stiffened skins) flat plate elements are inappropriate as the in-plane effects are generally important.

- When solving a problem by plate or shell elements never trust the results without making considerable effort to verify the correctness of the model as well as the solver algorithms and solutions.
- Depending on the properties of the material there will be a thickness value of a shell structure above which the model must have thick shell elements in preference to thin shell elements.
- When a model has thick plate elements or their shell element counterpart then the presence of 'locking' should be easy to recognize because the results will be low. Refining the mesh such that the aspect ratio of the length of an element side to the thickness is reduced should eradicate the detrimental effect of locking.
- The plate or shell theory is not appropriate in regions where panels join or curved surfaces make sharp radius turns.
- When using beam off-set elements and shell elements to model a stiffened skin structure there may be a significant error associated with the coupling of beam and shell elements, especially for coarse meshes.
- Avoid distorting the shape of plate and shell elements from their parent shape as they are highly sensitive to any type of distortion.

5.8.5   The Role of Compatibility

If elements are compatible internally and across their boundaries then, as the mesh is refined, the solution will convergence monotonically to the exact solution of the finite element method. Note that the exact elastic solution is never achieved because the number of nodes, or degrees of freedom, must be optimized to balance the errors due to discretization and round-off.

Remember that plate elements and general shell elements do not satisfy compatibility completely. In fact, complete compatibility requires the following:

- Elements must have the same order, although one can mix three-sided and four-sided elements.
- There must be connection between the corner nodes of neighbouring elements and, if present, continuity between the edge nodes of adjacent elements.

Special transition elements allow the element order to change in a mesh, but transition elements must not be used in regions where accurate stress values are required.

5.8.6   Elements to Model Contact

In this situation the following guidelines may be of use:

- Before developing a three-dimensional model for a problem with contact between different parts, check that the package has three-dimensional contact algorithms as these are a current topic of active research.

- If the finite element package has gap or interface elements to model contact between different parts of a structure, it is often a requirement to have adjacent nodes on the surfaces of the separate parts.
- If the package has elements to model the penetration of one surface by the surface of another part and push it back (known as sideline or contact patches) there is not usually the requirement to have adjacent nodes on the surfaces of the separate parts. Note that impact problems are handled in a different way.

5.9 DETERMINING A STRATEGY FOR DISTRIBUTING THE ELEMENTS

Once the list of useful elements has been decided, the analyst must determine how the elements should be distributed. At this stage it is probably sufficient to consider those areas that require large numbers of elements and those areas that can be modelled coarsely. Clearly, the choice of the mesh density is dictated by the element type that is to be used and the expected stress distribution throughout the structure. In the following sections some simple guidelines will be given.

5.9.1   Coarse or Fine Distributions

A coarse mesh may be used in cases where limiting the deflection is the principal design criterion for a complex three-dimensional solid structure where the analysis is global. Here, the geometry may be approximated by a coarse mesh of flat-faced solid elements.

Fine meshes are required in the following examples:

- When accurate stresses are necessary for a detailed failure analysis. Here, curved solid elements with minimum distortion may be used which fit the surface geometry as closely as is practical. In this case, the region of the structure that is of particular interest needs to be meshed very finely.
- When the effect of joints is modelled. Here the effect is often underestimated, but it may have an appreciable effect on the global, as well as the local, behaviour. In fact, it is much easier to devise a mesh for a complicated smooth three-dimensional structure than it is to model a joint accurately within a simple framework.

One of the main advantages of the finite element method over other numerical methods is that it allows a variable mesh density. Despite this, as a guide:

> The best results will always be obtained if the size of the elements in the mesh is constant and their shape is that of the parent element.

Often this is not possible and so the following guidelines may be of use:

- The volume of adjacent elements with identical material properties should not differ by more than two to four times. If there is an abrupt change of material properties the adjacent element volumes should be chosen such that the stiffness terms in their element stiffness matrices are similar.
- Avoid having elements of significantly different sizes in a model as this provokes ill-conditioning. As a rule of thumb the ratio of maximum to minimum element volume within a mesh should not exceed 30.
- The more gradual that any transition of material properties, loading, constraints or geometry can be made then the better the results will be. This is because the finite element method gives, in general, a smeared average solution with accurate results at only a limited number of points within each element. These points are not necessarily the nodes or the Gauss points.
- To minimize the problem of ill-conditioning when the element includes in-plane membrane and bending deformations it is advisable to have a mesh with a higher level of refinement than would be adequate if the element had either of these deformation components alone.

Distortion of elements can also cause problems as numerical errors are introduced. This is often the situation when the meshes and geometries are irregular. A list of the types of distortion is given in Chapter 3 and includes non-unity aspect ratio, angular distortion, volumetric distortion, warpage angle distortion and edge–node position distortion.

Packages check for distortion by determining the value of the Jacobian. It is often the analyst's responsibility to carry out checks to determine the maximum allowable distortions for any element.

5.9.2 Modelling Discontinuities

High rates of change in stress (and strain) exist wherever there is any type of discontinuity. For example, there can be discontinuities due to abrupt changes in such things as geometry, loading, material properties and supports.

As the mathematical formulation of elements tends to smear out the variation of displacement over an element, this has serious consequences when discontinuities are to be modelled. The finite element method is most accurate for continuum problems where the geometry, material properties and boundary conditions (loadings and constraints) all change in a smooth manner. If discontinuities of any form are present they are effectively smeared out and the finite element modelling is not precise. Refinement of the mesh helps to reduce problems, but doing this is too costly in three-dimensional problems, as to refine the mesh in all directions leads to an increased analysis cost which is roughly proportional to the cube of the number of elements.

Multi-point constraints can be used to allow abrupt changes in element geometry. This can be useful when mixing element types (e.g. two-dimensional membrane continuum elements and one-dimensional beam elements or three-dimensional solid continuum elements and general shell elements) as the analyst can ensure that the connecting degrees of freedom are matched by using multi-point constraint equations.

Once a choice of element or elements has been made and a distribution strategy decided, then the form of the mesh structure must be determined. By arranging the elements in different topologies, various forms of mesh can be built; this will be discussed in the next section.

5.10 TYPES OF MESH STRUCTURE

5.10.1 Regular and Irregular Mesh Structures

Now that the constituent parts of a mesh are known, the arrangement of these parts through the domain can be considered. This arrangement is known as the *form* or *topology* of the mesh. When using the finite element method, the points are the nodes of the set of elements used to split up the material volume and the elements can be arranged in any way, providing that the faces of the elements are positioned correctly. This means that, to ensure compatibility of the mesh, the edges of two-dimensional and the faces of three-dimensional elements which are touching must be in contact, with edge exactly matching edge or face exactly matching face and with node matching node. Clearly, one-dimensional elements are only placed from node to node. This less restrictive form is allowed as the calculation on any one element requires information from that element alone. The interaction between the elements takes place when the element equations are added together to form the global equations as discussed in Sec. 3.9.

From this it can be seen that there are two ways in which the mesh structure can be arranged. The first is a *regular form* or *topology*, where the nodes of the mesh can be imagined as a grid of points placed in a regular way throughout a cuboid. These nodes can then be stretched to fit a given geometry and this is shown in Fig. 5.3. Note that when the mesh is stretched the connections between the nodes do not change. The stretching takes place as if the mesh were made of rubber, and the topology of the mesh remains the same. Consequently, if any node in the mesh is considered it will be connected to the same neighbouring nodes both before and after the stretching process. Sometimes these meshes are called *structured meshes* as they have a well-defined structure, or *mapped meshes* as they can be seen as a cuboid mesh that has been mapped onto some other geometry. When considering these meshes it is useful to think of a *local coordinate system* within the mesh. This enables the orientation of the cells relative to each other to be determined, and so before the mesh is transformed the axes of this system are the edges of the cuboid. Once the transformation into the actual coordinate system, the *global coordinate system*, is carried out, then the local coordinate system axes become dependent on the position within the mesh. This is shown in Fig. 5.3.

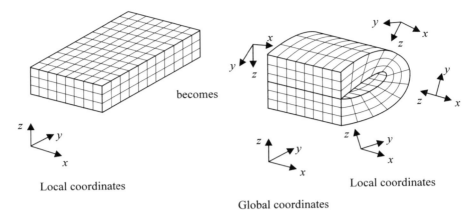

becomes

Local coordinates

Global coordinates

Local coordinates

Figure 5.3 Transformation of a mesh with a regular structure.

The second arrangement is an *irregular form* or *topology*, where the nodes fill the space to be considered but are not connected with a regular topology. Figure 5.4 shows a two-dimensional example of this type of mesh formed with triangular elements. Note that the element faces do not overlap and that elements are labelled by numbers and nodes by letters. It can be seen from

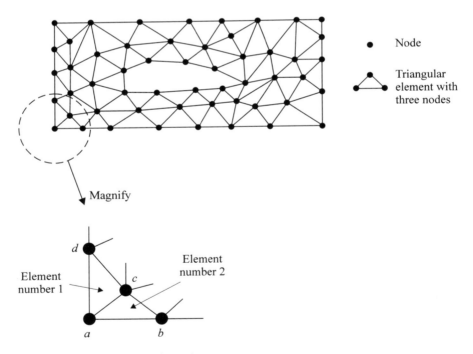

● Node

Triangular element with three nodes

Magnify

Element number 1

Element number 2

d

c

a   b

Figure 5.4 A mesh with an irregular structure.

the magnified section of the mesh that element number 1 has the three nodes labelled *a*, *c* and *d* at its corners, and that element number 2 has the nodes labelled *a*, *b* and *c* at its corners. The fact that any particular node is attached to an element cannot be known from the form of the mesh, and so a numerical table must exist that describes the arrangement of the mesh by listing which nodes are attached to each element. This contrasts with the regularly structured mesh where a knowledge of the location of an element within the mesh enables the labels of the points at its corners to be found implicitly. A mesh with an irregular structure is often referred to as an *unstructured mesh* or a *free mesh*.

Many structures that are of interest to engineers have complex geometries. With some ingenuity on the part of the analyst, it is possible to fit a mesh with a regular form to some of these geometries, but with many geometries this is not possible. This is where meshes with an irregular form can be used to great advantage, as these can be used to describe the most complex of geometries since there is no restriction on the form of the mesh. This can make the mesh generation process much easier and in some cases it is a prerequisite for producing a simulation. Another advantage of using irregularly structured meshes is that they can be created by automatic mesh generation algorithms, some of which are described in Sec. 5.11.3. These algorithms generate meshes that are unstructured using elements such as plane triangular elements and tetrahedral elements.

5.10.2   Determining the Mesh Form

Having chosen the element types and having made sure that the computer model of the geometry is complete, the next step is to decide the type of mesh form to be used. Often this is done by trying to find topological rectangles or bricks in the geometry in an attempt to see whether a regular mesh can be used. If this is not possible then an irregular mesh structure is appropriate.

Having decided on the mesh form, a mesh layout can be determined and an estimate made of the number of elements required. To do this requires considerable user experience; both the layout and number of elements depend on the strain (and stress) gradients that occur within the structure as it is deformed, as was seen in Chapter 3. In particular, the mesh layout depends on five things:

- the geometry of the structure
- the type of analysis, i.e. static, dynamic, thermal or nonlinear
- the boundary conditions
- the loadings
- the required results.

Remember that when constructing a mesh the elements must be arranged in such a way that the arrangement does not significantly affect the results. For example, with a regular mesh of triangular elements for membrane action it is good practice to have the direction of the sloping side alternating. The practice of having all the sloping sides in one direction is to be avoided.

5.11 BUILDING MESHES

5.11.1 Building Simple Meshes with a Regular Form

Many problems can be solved by using a mesh that has a simple regular form. One common way of producing a regular mesh is to use the hierarchy of entities as shown in Fig. 5.5, or something similar. In this figure the hierarchy for four-noded two-dimensional elements or eight-noded three-dimensional elements is considered. At the bottom of the hierarchy is the most basic geometrical entity which is a point, several of which can be linked to form lines (from two points), arcs (from three points) or splines (from four points or more). By combining adjoining lines, arcs and splines the third-level entity, the edge, can be created. If four edges form a closed loop they can be seen to be the boundaries of a surface and six of these surfaces can be used to bound a volume. This set of relationships is determined by the form of the elements being considered. Once the surfaces for a two-dimensional problem or the volumes for a three-dimensional problem are defined, the elements can be formed. This is done by mapping the surfaces into a square, and by mapping the volumes into a cube. These squares and cubes are used to define a local coordinate system in which the parent elements can be created before these are transformed back to the global coordinate system in which the real elements are defined. Here, the techniques of transformation that are used are similar to those used in forming the isoparametric elements discussed in Sec. 3.7.

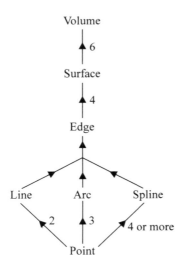

Figure 5.5 A hierarchy of entities.

Another way of building regular meshes in a topologically rectangular domain is to use the techniques of Thompson, Warsi and Mastin (1982), where partial differential equations are developed that describe the nodal positions within the domain. These techniques generate meshes which are particularly smooth in their variation from element to element.

5.11.2 Using Commercial Pre-processors

Commercial mesh-generation packages have been around for some time now aimed at the finite element structural analysis market. These packages usually have the following components:

- A geometry creation routine, where one-, two- or three-dimensional geometrical data can be created in the form of points, lines, arcs, splines and, sometimes, surfaces. An interface to extract similar data from CAD systems is a common feature as well.
- A domain definition routine. This allows the creation of surfaces, in two dimensions, or volumes, in three dimensions. These domains are the spatial entities on which the mesh is to be built.
- A mapped-mesh generation routine. This enables a mesh with a regular structure to be created within the domains. These domains must be topologically consistent with the element type being used. For example, if four-noded quadrilaterals are being used to mesh a two-dimensional domain, then a four-edged domain must be used.
- A free-mesh generation routine. This enables a mesh without a regular structure to be created within the domain. In this case there is, in principle, no restriction on the form of the domains, and so they can be either surfaces bounded by any number of edges (for two-dimensional problems) or volumes enclosed by a set of these surfaces (for three-dimensional problems).

When using commercial mesh-generation software, hierarchies such as that shown in Fig. 5.5 are used. Usually, this does not cause a problem, but there is one area where errors in the modelling of a geometry can occur. Coons patches are an obvious choice for defining the geometry of a surface within the hierarchy, as the edges are used to define the surface. As was discussed in Sec. 5.4, such a representation of a surface may not be adequate if the patch is too large for the curvature of the surface. One way of overcoming this problem is to define smaller surfaces, but this involves much more work on the part of the analyst. Another way is to use more accurate surface descriptions, say Bézier surfaces or NURBS, derived from a CAD model of an object. Many commercial finite element pre-processors can read these more accurate surfaces from the database of a CAD system. Then, a set of edges can be used to define a Coons patch surface. Once this has been done, the user can tell the pre-processor to calculate the mesh points on this surface by first calculating the coordinates of the points on the Coons patch and then recalculating the coordinates so that they are positioned on the exact surface.

5.11.3 Automatic Mesh Generation Algorithms

For simple geometries it is easy to see how a mesh can be built, but when the geometry becomes more complicated the meshing process is more difficult. Several techniques have been developed that can take complex two- and three-dimensional geometries and then automatically produce a mesh that models the geometry. Typically, the mesh will have an irregular structure. As was stated, generation is a costly part of the analysis process because of the large amount of human effort that can be required to build the mesh for a complex geometry. Any savings in the time taken to build a mesh reduce the cost of an analysis, and so these automatic mesh generation techniques are being actively researched (George, 1991).

The first method that will be discussed is *Delaunay triangulation* (Cavendish, Field and Frey, 1985; Holmes and Lanson, 1986; Watson, 1981). Figure 5.6 shows this algorithm at work for a two-dimensional case where triangular elements are to be created. The algorithm is easily extended to three dimensions where tetrahedral elements are formed. Figure 5.6(a) shows that the basic technique is started by producing nodes on the boundary of the domain and nodes inside the boundary. In this case there are 12 nodes on a square boundary and one node inside the domain. To ensure that the final triangulated mesh has no gaps in it, three extra nodes are then created that define a super-triangle. From Fig. 5.6(b) it can be seen that these extra nodes have to be placed so that they define a super-triangle that encloses all of the original nodes of the problem. This super-triangle is taken to be the first element and then one of the original nodes is used to split this element into three new elements (Fig. 5.6c). Now an iterative element creation procedure can begin. One by one each of the remaining nodes is considered and the mesh modified. To do this a circle is created for each element such that it passes through each of the three nodes of that element. Figure 5.6(d) shows the circles of the elements and node 2 will be considered. This node lies outside two of the circles and inside the other. The triangulation algorithm states that if a node lies inside a circle then the element to which the circle is attached should be deleted. Once all the necessary elements have been deleted, new elements can be created that include the node being considered. This is shown in Fig. 5.6(e) where the lower element of Fig. 5.6(c) has been deleted and three new elements have been created which are joined at node 2. Then another node is considered and the process continues. Eventually, a final mesh is created such as that in Fig. 5.6(f). This can be modified so that only the original domain, in this case the square, is modelled. This is done by deleting all the elements which are attached to the nodes that formed the super-triangle. Finally, the shape of the remaining elements is checked and, where necessary, the mesh is modified using face or edge swapping and node insertion to give elements as near to equilateral triangles as possible. At the end of this process the mesh does not have elements with a very distorted shape as these elements lead to numerical errors when calculating the element equations as discussed in Sec. 3.7.6.

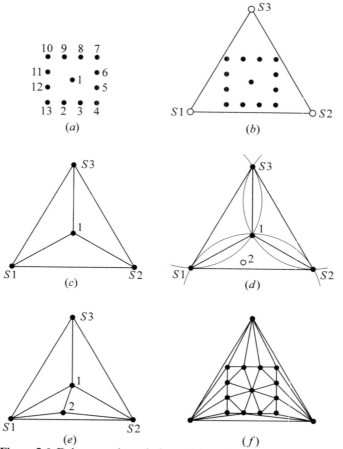

Figure 5.6 Delaunay triangulation: (*a*) boundary and internal nodes, (*b*) addition of super-triangle, (*c*) the first three elements, (*d*) forming circumcircles, (*e*) element deletion and recreation, (*f*) the final mesh.

The second method is based on the use of the *quadtree* and *octree* methods (Yerry and Shephard, 1984; Cheng, Finnigan, Hathaway *et al.*, 1988). These methods take a domain and place it inside four squares, if it is a two-dimensional problem, or eight cubes, for a three-dimensional problem. These are then sub-divided until the required definition is acquired. Hence the name 'quadtree' refers to the structure of the elements in two dimensions and 'octree' refers to the three-dimensional method. Figure 5.7(a) shows an example of a two-dimensional domain that is to be meshed. Four squares are placed over the domain, as shown in Fig. 5.7(b), and a node is created where the squares are joined inside the domain. Each square can then be subdivided into four more squares and more internal nodes created. Two further subdivisions are shown in parts (c) and (d).

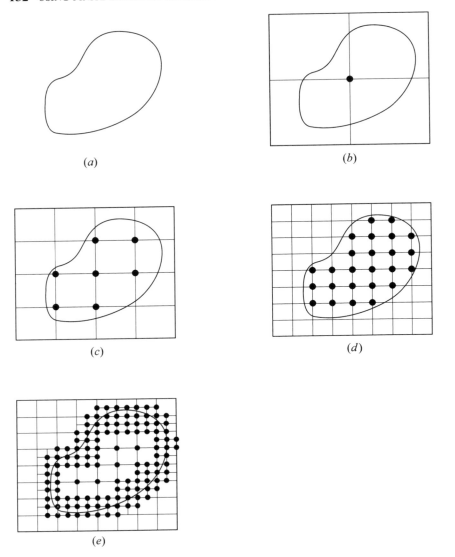

Figure 5.7 The quadtree method: (*a*) domain to be meshed, (*b*) first overlay, (*c*) second overlay, (*d*) third overlay, (*e*) selective subdivision at boundaries.

Once the element size for the bulk of the mesh is small enough, only the elements that cover the domain boundary are subdivided. This selective subdivision is shown in Fig. 5.7(e), and it can be continued as required. Note, however, that it leaves a mesh that is a stepped representation of the domain. To overcome this the external nodes are moved so that they are on the surface of the domain. Finally, triangular elements can be created to link all the nodes.

A third method that is increasingly popular is the *advancing front method*. This is described in detail by George (1991). Essentially, the method requires a mesh on all of the surfaces of the boundaries of the structure. This consists of line elements in two dimensions or triangular elements in three dimensions. These form a so-called *initial front* on which new elements, triangles in two dimensions or tetrahedra in three dimensions, are built. Where possible, equilateral triangles are used either on their own or to build the tetrahedra, so ensuring that the mesh has optimal shape properties. Once the first set of elements has been added the new front is established by removing the old face elements from the front list and adding the new faces as appropriate. Elements are added in this way until the whole volume is filled and the front has no members. Figure 5.8 shows the process in action.

Specialized software is available to perform mesh generation using forms of Delaunay triangulation, quadtree/octree and advancing front methods, but commercial finite element mesh generation software can also be used to generate a mesh with an irregular structure in an automatic way. This is often done by meshing the surface of the domain, using triangular elements. Then the volumes that have been defined by the surfaces can be meshed using tetrahedral elements formed from the elements on the surface. At first sight this might appear to restrict such free-mesh generation methods to using only tetrahedra. These can, however, be easily converted to eight-noded brick elements as shown in Fig. 5.9. There a single tetrahedral element is taken and new nodes formed at each of the mid-sides of the element edges, at the centroids of each face of the element and at the centroid of the whole element. These can then be joined as shown to produce four eight-noded brick elements.

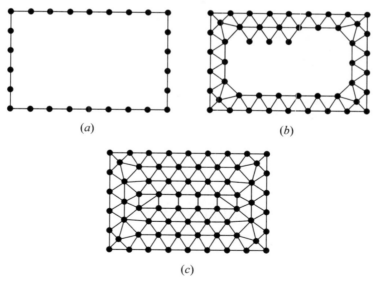

(a)   (b)

(c)

Figure 5.8 Mesh generation using the advancing front method: (*a*) the domain boundary (first front), (*b*) front advancing into domain, (*c*) final mesh.

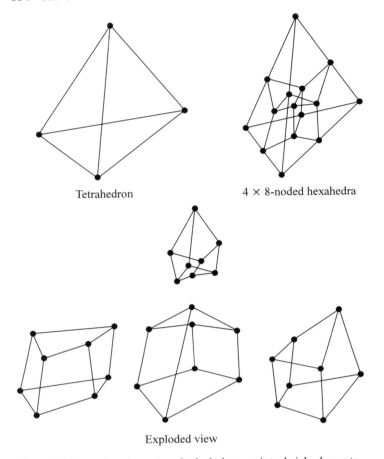

Tetrahedron   4 × 8-noded hexahedra

Exploded view

Figure 5.9 Transforming a tetrahedral element into brick elements.

Note that a three-dimensional mesh is very much more difficult to visualize when it is generated using an automatic routine. In this case mesh checking becomes a more important stage in the model development, as much of the model and many of the element edges and nodes are hidden from view in any graphical display.

5.12 ADAPTING A MESH TO IMPROVE A SOLUTION

5.12.1 Refinement and Enrichment

Once a mesh has been built it is possible to modify it in such a way that the solution that is produced on the modified mesh is a better one. This modification can take place before a solution to the structural problem is found, using an analyst's intuition to improve the mesh. Equally, it can take place after a solution is found, using either intuition again or some automatic means to refine the mesh such that a new and improved solution can be generated.

On many occasions, the solution will be good over most of the model but will need more refinement in one or two regions. On other occasions the model may be highly complex with many components such that refinements in all the components soon become impractical. This situation calls for the ability to model a subregion separately while including the influence from the overall structure.

Mesh modification techniques can be applied after a solution has been produced on an initial mesh. These techniques are used to modify the mesh in the light of the results achieved on it and so the dependence of the quality of the results on the user's experience is reduced. These modification procedures require that an initial analysis is made using a crude but realistic mesh of the domain. From the results of this initial analysis the mesh is recreated such that the density of the mesh points is greatest in regions of the domain where the strain gradients change rapidly or where the error in the numerical equations is found to be large (Zienkiewicz and Taylor, 1989). The mesh is said to be adapted to take account of the results generated. Several types of mesh modification are commonly used:

- *Mesh enrichment,* where additional points are placed within the domain at the locations where they are needed as shown in Fig. 5.10. In this figure a mesh is required for a model of a plate which has a hole in it, leading to a stress concentration when it is loaded in the plane. The original mesh of triangles has a regular spacing but the enriched mesh has additional nodes and elements in it so that there are more elements near the stress concentration. This technique is usually applied to meshes that consist of triangular elements in two dimensions and tetrahedral elements in three dimensions. Such meshes allow additional points to be created in the mesh and then the Delaunay triangulation method, or similar methods, can be used to create a new set of elements.

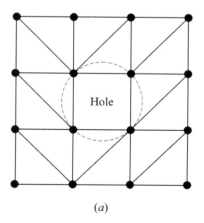
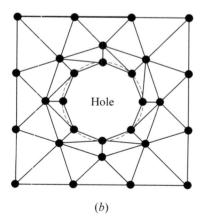

(a)   (b)

Figure 5.10 Mesh enrichment: (a) original mesh, (b) enriched mesh

- *Mesh adaption,* where the topology of the mesh stays the same but the mesh points are moved so that the density of points increases where required as shown in Fig. 5.11. Here, a stress concentration is again modelled. Note that the number of nodes and elements remains the same in the adapted mesh. Only the node positions are changed. This movement of the points can be brought about by using modified forms of the partial differential equations that are used in some grid generation methods as was discussed in Sec. 5.11.1.
- *h-refinement,* where a mesh is refined by systematically subdividing elements into more elements of the same order. For example, a quadrilateral element could be converted into four quadrilateral elements placed within the same area. On each h-refinement the previous mesh form can be seen within the new refined mesh form.
- *p-refinement,* where the elements are modified to be elements covering the same domain but with an increased order of polynomial interpolation. For example, linear elements could be converted to quadratic elements then cubic elements and so on.
- *hp-refinement,* where both of the above techniques are used together.

By using these smoothing or adaption techniques the accuracy of the solution can be increased, but there is a penalty in that extra computational effort is required. In some commercial packages adaptive meshing is implemented as a way of achieving automatic convergence of finite elements through mesh refinement or remeshing a model based on error estimates.

5.12.2 Reducing the Bandwidth

As mentioned in the Sec. 3.9, the use of direct solvers for many structural finite element solutions leads to a solution time proportional to the number of variables times the half-bandwidth squared. For the sparse matrices that are produced the half-bandwidth is a function of the way in which the nodes of the mesh are numbered.

For a simple mesh such as that shown in Fig. 5.12, it is easy to see that different ways of numbering the nodes can be chosen. The mesh on the left has

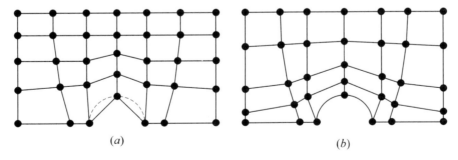

(a)   (b)

Figure 5.11 Mesh adaption: (a) original mesh, (b) adapted mesh.

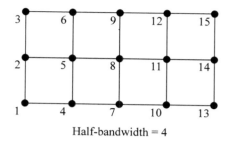 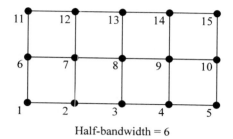

Half-bandwidth = 4   Half-bandwidth = 6

Figure 5.12 Effect of node numbering on half-bandwidth. All elements are four-noded quadrilaterals.

a bandwidth of four while that on the right has a bandwidth of six. Clearly, the analyst can influence the size of the bandwidth during mesh preparation and so it is common for manual methods of bandwidth reduction to be built into mesh creation strategies. These are discussed by NAFEMS (1986) and Desai and Abel (1972).

For complex meshes, however, manual intervention in the node numbering problem for minimum bandwidth is virtually impossible. Hinton and Owen (1979) discuss this and show an example of an automatic procedure to reduce the bandwidth of a mesh. This is based on the method of Cuthill and McKee (1969) which utilizes the quantity known as the degree of a node, i.e. the number of nodes connected to a given node, to reduce the bandwidth. The method is illustrated in Fig. 5.13 by a flowchart. Note that a start node must be chosen by the user and is assigned as node 1. Then all nodes attached to this node are numbered in sequence 2, 3, 4 and so on in increasing order of degree. Then the nodes attached to node 2 are renumbered in the same way. This is repeated for all nodes in the mesh. There is no guarantee that this will be an optimal numbering and so it is common for the algorithm to be repeated with a variety of nodes being taken as the first.

Figure 5.14(a) on page 139 shows a mesh of triangular and quadrilateral linear elements together with the degrees of each node. Two different solutions obtained using the Cuthill–McKee algorithm are also shown in parts (b) and (c). Note that the bandwidth is increased from 7 in part (b) to 8 in part (c).

Clearly, several attempts have to be made as different start positions will give different values for the bandwidth. The Gibbs, Poole and Stockmeyer (1976) algorithm addresses this problem by finding a start node with the minimum degree on the boundary of the domain and finding an optimum path from the start node. This is also discussed in George (1991) together with a variety of other methods.

5.13   CHECKING A MESH

Once a mesh has been built it is difficult to check the data for errors. To assist in this process various techniques can be used and most are built in to commercial pre- and post-processors.

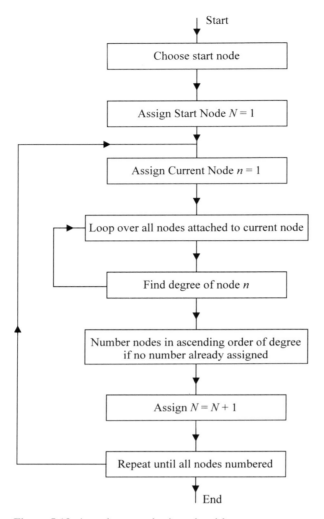

Figure 5.13 A node renumbering algorithm.

The following checks are common:

- *Finding coincident nodes* A tolerance is given and the positions of all nodes are checked against all other nodes for coincidence. Any sets of nodes that are listed as coincident can usually be merged together by the software and the appropriate changes made to the connectivity list automatically.
- *Free-face checks* The faces of the elements that are only connected to one element are displayed. This is done using the algorithm given in Chapter 6. All free-faces should be on the true boundary of the mesh and so this method shows any holes in the mesh that have been created inadvertently.

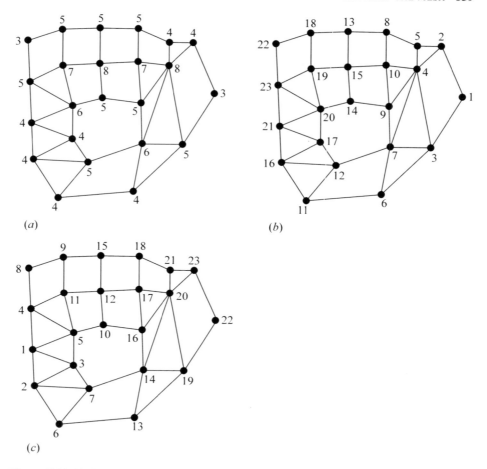

Figure 5.14 Node renumbering with the Cuthill–McKee algorithm: (*a*) the degree of the nodes, (*b*) case 1—bandwidth of 7, (*c*) case 2—bandwidth of 8.

- *Free-edge checks* This is similar to a free-face check but the free edges of the mesh are displayed. Again holes in the mesh, created accidentally, are highlighted.
- *Exploded mesh plots* The elements are drawn on the computer screen in a shrunken format. This aids visualization of the mesh.

6

OBTAINING AND ANALYSING THE RESULTS

This chapter contains material previously published in Shaw (1992), by kind permission of Prentice Hall.

Once a mesh has been built to describe the domain occupied by the structure, the rest of the computer model can be built. It is only at this stage that the description of the physical problem generated in the initial stage of the analysis process can be related to the computational geometry described by the mesh of nodes and elements. For each element, its material properties must be defined together with the boundary conditions on the faces of the elements, or at the nodes, which form the exterior of the mesh. The first few sections of this chapter will look at how this can be achieved for the case of a linear static analysis. Be aware, however, that for more complex problems, such as those discussed in Chapter 10 where optimization, dynamics, nonlinearity, time dependence and thermal effects are involved, much more information must be provided at this stage.

Once the computer model is complete, the solver can be run to obtain the results of the simulation. While this should, in theory, be a straightforward process for a linear static analysis, things can and do go wrong. The middle sections of this chapter look at how the solver is run and the troubleshooting that is required to ensure that results are produced. These results will be a numerical solution to the governing equations, produced with the appropriate boundary conditions on a mesh that approximates the geometry of the problem. Hence, the solution is strictly a solution of the numerical problem and not of the physical problem, any differences between these two being due to such things as an inadequate mesh or an approximate boundary condition specification.

When the numerical solution is obtained it is necessary to determine whether or not it bears some relationship to the physical reality. If it is likely that it does, then the required technical information can be extracted from the results with some confidence. Consequently, the final sections of this chapter look at what forms the results of a simulation, how computer graphics can be used to start the evaluation of the results, how the solution can be checked to see if it is likely to be reliable, how the model can be refined so that the required data can be obtained from the results and at failure criteria.

The issues discussed in this chapter will be amplified by the discussion of the examples of finite element modelling in Chapters 7–9.

6.1 SPECIFYING MATERIAL PROPERTIES

As was seen in Sec. 2.2.6 it is not necessarily a straightforward task to define precisely the material properties and, frequently, they must be approximated when compiling the model data for an analysis. Typical property data in Table 2.1 shows that there is a degree of variation of the Young's modulus and strengths for most engineering materials. It is therefore the responsibility of the analyst to choose the appropriate value for the properties that ensures the analysis provides results for a conservative and safe structural design.

Note that data in tables, such as Table 2.1, are usually quoted as being valid at room temperature (20 °C). Hence, if the operating temperature is not room temperature, a check needs to be made on how much a property will vary between the temperatures. Furthermore, if there is a temperature variation throughout a body which is made of notionally isotropic material then the material must be considered as nonhomogeneous as the material properties will be dependent on both position and temperature. Many analysis packages have the facility to model the variation of a property with temperature and will assign a constant 'smeared' property value to each element in the mesh.

There are other occasions when the analyst is required to model real material behaviour using a smeared equivalent material property. One important application of this approach is in the analysis of a composite material that consists of two or more distinct solid phases with one of the phases having constant shape and size and being distributed uniformly throughout the second phase. There are in the literature standard techniques for calculating equivalent material properties from knowledge of the constituent properties and the distribution and volume fraction of the dispersed phase. For this approach to work the dimensions of the composite material in the structure must be greater than 20 times the smallest dimension of the inclusions. This is the situation, for example, with fibre-reinforced composite materials that are increasingly preferred to conventional materials in major structural applications.

On a much larger scale there are many occasions when part of a structure contains closely spaced and regularly repeated geometrical features; a perforated boiler tube plate being an example that can be visualized. In such cases, it is not sensible to model all the holes individually, as this would require many elements. Nor would it be correct to ignore their presence, as they have a significant effect on the structural behaviour of the plate. The solution to this problem is to model the plate as a single continuum with equivalent elastic constants that take account of the holes. A finite element modelling procedure can be used to determine these equivalent properties (NAFEMS, 1986). Having obtained a solution, by modelling the plate model as a continuum, the results are used to obtain stresses at the local level by analysing separately a hole and a representative surrounding region of the plate. A similar procedure can also be used to model corrugated plates and bolt flanges that contain a series of bolt holes around the circumference.

Finally, note that even if the finite element mesh and the boundary conditions faithfully reproduce the real physical situation being modelled, there could well be some uncertainty in the property data of the material or materials of the structure, and that this often limits the accuracy of the solution. Hence, the quality of the material property data can have a direct influence on how much effort the analyst needs to spend in producing an accurate finite element model.

6.2 FINDING THE BOUNDARIES OF THE MESH

When calculating the required solution, the governing partial differential equations are solved subject to the appropriate boundary conditions. During the specification of the physical problem the boundaries are defined in terms of the geometry of the structure, but now these boundaries must be found in terms of the mesh that is being used to describe the structure. This involves defining the boundaries as a collection of element faces before, possibly, listing the boundary nodes.

6.2.1 Boundaries of Meshes with a Regular Structure

If the mesh has a regular structure, a knowledge of the local coordinate system (see Sec. 5.10.1) can be used to define a set of indices i, j, k. These indices denote the position of an element within the mesh structure and range from unity to the maximum number of elements in each of the local coordinate directions. The local coordinate system can also be used to define the faces of an element within the mesh. Figure 6.1 shows mesh with a regular structure shown in terms of its local coordinate system. Each element of the mesh has six faces and a typical element is shown with its faces labelled with the points of a compass. Hence the faces are named North (N), South (S), West (W), East (E), Top (T) and Bottom (B). These latter two are also known as High (H) and Low (L) in some programs. For example, the face of the element that is at the most positive local x-direction position, in the direction of increasing the index i, is the East face and the one at the most negative local x-direction position is the West face.

It can also be seen, by looking at Fig. 6.1, that any plane of elements has a constant value of either i, j or k, and that the extent of the plane can be defined by knowing the limits of the other two indices. The patch of elements shown in Fig. 6.2 has a constant value of the index i and the limits are defined by j_{min}, j_{max}, k_{min}

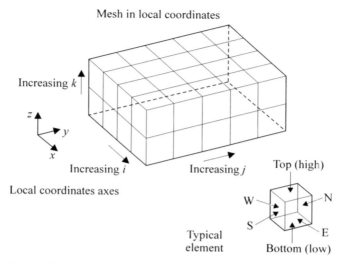

Figure 6.1 Use of a local coordinate system.

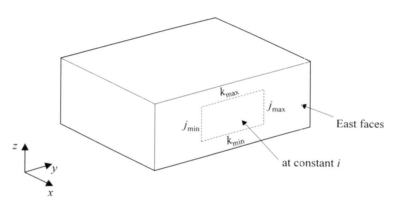

Figure 6.2 Defining a patch of cell faces.

and k_{max}. Also, the faces of the elements in the patch shown are in the positive local x-direction and so they are all East faces. By using this notation a set of patches can be defined on the boundaries of the mesh. These patches have to be defined for all the surfaces where the boundary conditions are to be specified.

Remember that when defining a patch of element faces on a boundary, it is sensible to define patches that have only one boundary condition type applied on the patch. This means that the faces or nodes of the whole patch might be, say, restrained in some direction or have a constant pressure load applied. By doing this it is simple to specify the boundary condition that applies on a patch by a single command.

6.2.2 Boundaries of Meshes with an Irregular Structure

When a mesh has an irregular structure the problem of defining the boundaries becomes much more difficult. Actually finding the element faces that are the boundaries of the mesh is quite straightforward, as will be seen. It is the collecting of the various element faces into groups that are suitable for the addition of the same boundary condition that is difficult.

Two pieces of information help us to find the element faces that are on the boundary of the mesh. First, each face of an element is uniquely defined by the nodes that are on the face and, second, the faces on the boundary of the mesh can be associated with only one element, while those internal to the mesh must be associated with two or more elements. This is shown in Fig. 6.3, where it is clear that the internal face is common to the two elements and that the external faces are related to only one of the two elements.

The process of finding the faces that are on the boundary of a mesh is called a *free-face check*. The algorithm used to do this is shown in Fig. 6.4, from which it can be seen that each element is considered in turn. Then each face within an element is found in terms of the numbers of the nodes attached to it. A unique label for each face on the element is found from these node numbers. Each of these face labels is checked against a list of the face labels stored in a database. This database is created as the process is carried out and records the number of elements to which a given face is attached. If a face label does not exist in the database then an entry recording the new face label is made in the database and the count of occurrences of the face is set to unity. If the face has been listed before, the count is increased so that it reflects the number of elements associated with the particular face. Once all the faces on a element have been processed then a new element is chosen, and after all the elements have been processed the database is complete. By checking the database, a list can be made of all those faces that are attached to only one element. These must be the faces on the boundary of the mesh, and the list of faces is known as a *free-face list*.

Once the free faces have been identified, they can be grouped into the required sets of faces for the different types of boundary conditions. This is usually done by

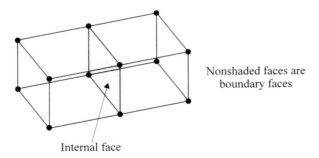

Nonshaded faces are boundary faces

Internal face

Figure 6.3 Internal and boundary faces.

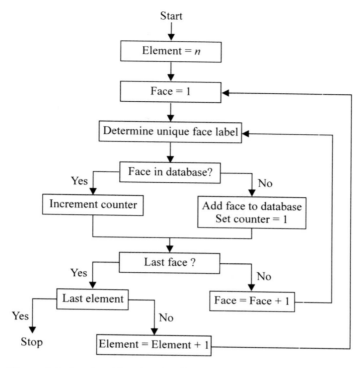

Figure 6.4 An algorithm for the determination of free-faces.

displaying the faces in the free-face list on a graphics screen in a variety of ways, including the following:

- A hidden-line display, where the user sees the faces just as they would be seen if they existed physically. That is, faces that are behind other faces, as seen by the viewer, are hidden from view.
- A display of the faces within a given volume.

Once the displays of the bounding faces of the mesh have been produced the pointing device of the terminal or workstation can be used to pick out the faces. Either this can be done face by face, or whole sets of faces can be picked by placing a window on the screen and noting the faces that are within the window. This is illustrated in Fig. 6.5 which shows a simple mesh made from two mesh blocks of regular structure. Here it is desired that a constant pressure boundary condition is applied at the nine element faces labelled with arrows in the left-hand view. These faces may be picked manually using the cursor on the display screen, but, by changing the view of the mesh to that shown on the right-hand side of the figure, a rectangular window can be defined using two corner points as shown. Then all the faces that are wholly within the window, the nine required faces, can be labelled by the pre-processor as being a set of boundary faces. This windowing

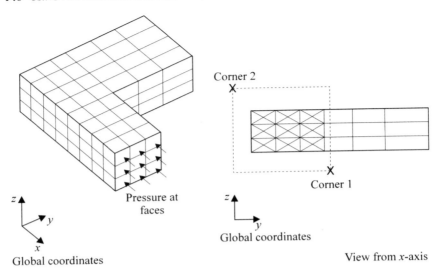

Figure 6.5 Finding boundary faces on the screen.

method has great advantages when dealing with large numbers of boundary faces.

6.2.3 Grouping Faces at the Boundary Together

Regardless of whether the mesh has a regular or an irregular structure, the boundary faces must be grouped together into sets of faces using the methods just described. Each set of faces can then be given an index that allows the set to be related to a boundary condition. Sometimes, the boundary condition on a set of faces is unique to that set, however, in some cases, the same boundary condition may well be applied to several sets of faces. In this latter case, each of the sets can be given the same index and then the index can be linked to the given boundary condition.

6.3 IMPOSING THE BOUNDARY CONDITIONS ON THE MESH

Once the boundaries of the mesh have been identified the appropriate boundary conditions can be applied at these boundaries. For a linear static analysis the displacements of the nodes resulting from a series of forces applied at the nodes are calculated. Hence it is obvious that a series of consistent nodal forces must be specified, which can be done either directly or indirectly as a pressure loading. Less obvious is the requirement that some displacements must also be specified. These must at least restrain the structure from rigid-body motion where the whole structure has a constant load applied to it and accelerates uniformly through space.

One rule that must be kept in mind is that, at the nodes where restraints are applied, the associated nodal forces cannot also be applied as the numerical solution will overwrite the values of the force at these nodes to ensure that the correct displacement is calculated.

6.3.1 Point Forces

In the real world forces are applied to a structure through a contact patch on the structure. Even when the loads are effectively point loads this patch must have some finite area and so the modelling in the finite element analysis is an approximation to reality. Hence the analyst must consider in detail how to represent the physical problem through the finite element model.

To represent even point forces, therefore, the appropriate load might have to be applied over several nodes depending on the size of the elements in the region of the contact. For example, Fig. 6.6(a) shows the physical loading of a two-dimensional structure, together with two ways in which the load could be modelled. In Fig. 6.6(b), the mesh is sufficiently fine in the region where the

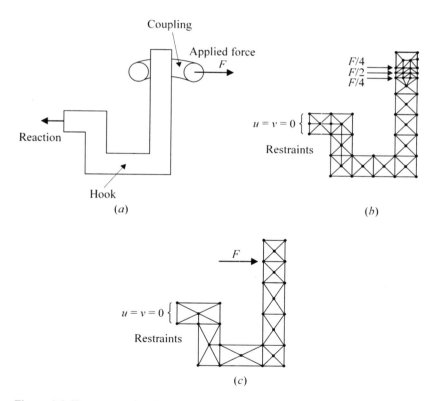

Figure 6.6 Forces on a hook: (*a*) the situation, (*b*) with a fine mesh, (*c*) with a coarse mesh.

force is applied and so the load has been apportioned over three nodes, whereas in Fig. 6.6(c) the mesh is coarse and the load has been applied at just one node. In a similar way distributed loads can also be applied over a surface. For example, a constant pressure load may be represented by an averaged force at a series of nodes.

Clearly, for both two- and three-dimensional problems, the forces will be defined using appropriate components in, say, the x-, y- and z-directions.

6.3.2 Pressure Loads

On element faces or edges on a boundary pressure loads can be applied by most programs. These require just the pressure on the element face or edge to be specified. From the element geometry, the face area or edge line for a plate or shell element can be found and hence appropriate nodal loads can be applied on the element face or edge, or on a series of linked element faces or edges.

It is known that modelling errors are introduced, when applying pressure loads, if the real geometry differs from that of the mesh and the elements are distorted from their parent shape.

6.3.3 Fixed Displacements

To prevent the structure from moving freely in space, which is a numerically singular problem, it must be restrained in all the coordinate directions at one node at least. This means that a displacement of zero must be specified in all the coordinate directions for one node. Sometimes whole sets of nodes need to be restrained in all directions to simulate a rigid fixing such as a floor mounting. An example of this is the restraint applied to the left-hand end of the hook shown in Fig. 6.6. There the nodes are restrained by setting the displacements in both coordinate directions to zero. Implicitly, the displacement in the third coordinate direction is set to zero for a two-dimensional program, but when a three-dimensional program is used the analyst must remember to set the displacement to zero explicitly.

A common use of fixed displacements is in the simulation of sliding, where nodes are free to move in one direction but are restrained to have zero movement in all other directions. This is shown in Fig. 6.7 where sliding of the arm in the vertical direction is allowed. Here the z-direction is the vertical direction and so the displacement of the upper end of the arm in both x- and y- directions is set to zero. Equally all three displacements at the pinjoint are set to zero.

6.3.4 Fixed Rotations

As well as fixing displacements, the rotation at some nodes may also have to be specified. Examples of this will be given in the examples in Chapters 7–9.

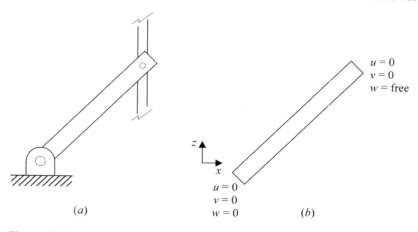

Figure 6.7 Example of applying restraints: (*a*) the situation, (*b*) the restraints applied to model the situation.

6.4 CONTROLLING THE SOLUTION

At this stage in the analysis process the model data consists of the finite element mesh defined by the nodal coordinates and the element connectivity, the material properties for the elements and the boundary conditions in terms of forces and displacements at appropriate nodes. Whether this data is stored in the database of a pre-processor or as a collection of records in a neutral (ASCII) file it must now be made available for the solver to read. For a linear static analysis the only other parameters that the solver might need are the method of solving the linear equations and any associated parameters. For example the solver may have to be instructed to use a direct solver or iterative solver of a particular type. In the former case no further information is required but in the latter case the number of iterations and any relaxation parameters need to be specified.

For more complex analyses such as those discussed in Chapter 10 other parameters such as the initial conditions, the number of iterations to remove nonlinearity, stress or displacement constraints for optimization problems and the method of eigenvalue analysis are required. Finally, in all cases, file management data is needed to specify the source of the input data for the solver and the destination of the output data from the solver. One useful file handling device that can be employed is to store the upper and lower matrices, decomposed from the global stiffness matrix, on disk. This enables further simulations using the same mesh, with say different loading conditions, to be run without the computational expense of recalculating these matrices.

6.5 RUNNING THE SOLVER

Once all of the model data and management data have been assembled, it is necessary to issue the command that runs the solver program. This can be done manually, once all the input files are in the correct place in the directory structure of the correct computer, but many applications now make use of a simple command file written in the operating system language of the machine where the pre- and post-processing takes place. This command file performs the following functions:

- It copies the input data for the solver to the correct directory (and correct computer if necessary).
- It tells the solver where the find the input data.
- It runs the solver program.
- It tells the solver where to write its output data.
- It copies the output data, both results and management data, to the correct directory (and correct computer if necessary).

Using such files the process of running the solver becomes extremely simple for the analyst.

6.6 TROUBLESHOOTING

Once the solver has run the results should be available for the analyst to check. However, it is very common for analysts to check for results and find none available, perhaps for the following reasons:

- The solver has not finished its calculations.
- The solver has encountered an error when reading the input data, such as the input data files not being in the correct location or the data files being in an incorrect format.
- The solver finds that no solution can be found to the linear simultaneous equations owing to incorrect data such as a mesh with coincident nodes or collapsed elements or inadequate restraints on the boundary.
- The solver encounters an error when writing the output data, for example when the target disk for the output data is full or the target directory does not exist.

Consequently an analyst should do the following:

- Check to see that the solver has finished by interrogating the computer's list of processes running at any one time.
- Look at any error and warning files written by the solver to see if there has been a failure. It is here that both file management errors such as incorrect directory specification or disk full errors as well as errors in solution will be found. Also the solver may write warning messages to alert the user to potential inaccuracies.

- Check and, if necessary, correct the command file.
- Check the integrity of the mesh in terms of coincident nodes and spurious free faces and modify where necessary.
- Check the material and physical properties of the elements, and modify where necessary.
- Check the boundary conditions to ensure that the correct forces in the correct directions are applied and that the structure is adequately restrained to remove numerical singularities due to rigid-body motion.

When all these have been done, and any errors corrected, it should be possible to resubmit the command file and find that on completion the solver has produced the necessary results files. Often this does not happen and so the checking process might need to be repeated and further corrections made. Even when results are produced they may not be of any use and determining whether this is the case or not is the subject of the next few sections.

6.7 THE RESULTS GENERATED BY THE SOLVER

When the solver runs it produces a large amount of data that has to be analysed. This analysis is undertaken so that the quality of the solution can be examined and so that useful technical information can be extracted where appropriate. First, it must be decided what information will actually be available at the beginning of the results analysis.

Information can be produced by the solver in two main forms. These forms differ in how the data is stored by the computer. In one form the data is stored using an internationally agreed format that defines individual characters of data such as the letters of the alphabet or the numbers 0 to 9. This form of data is known as *ASCII data*, after the committee that divised the data standard, and can be written to a terminal screen or stored in a file known as an *ASCII file*. Each character has to be defined by one byte, i.e. 8 bits, of computer memory and so 256 different characters can be specified. ASCII files of data can be edited by text processors and other software, and they are effectively machine independent which means that the data can be transferred from one computer to another computer, even if the machines are from different manufacturers, without any translation process taking place. While most computer manufacturers use the ASCII standard, there are other standards such as EBCDIC which are used by a minority of manufacturers.

Numerical data can also be stored in the second data storage format, which is known as *binary data format*. There is a standard for this method of data storage, but usually, at present at least, the method of storage is peculiar to each computer operating system or computer manufacturer. Each of these binary storage methods enables real numbers, for example, to be stored by four bytes in single precision or eight bytes in double precision. Binary data is stored in files known as *binary files*. These files are not machine independent and so cannot be transferred from computer to computer without some form of translation process taking place. Sometimes when a workstation, for example, is connected to a minisupercomputer a translation program is provided by the workstation vendor to facilitate the transfer process. By using binary files to store real numbers, there is a saving in the amount of storage required, as can be seen from the number of bytes required to store each number.

The type of information produced by the solver program can usually be controlled by the user but it often consists of the following:

- Values of the residual error if the linear equations have been solved iteratively. This gives a measure of the accuracy of the solution in numerical terms on the mesh used.
- A complete list of the variables at all the nodes of the mesh or all the elements of the mesh. Typically the list consists of nodal displacements, and stresses and strains at the nodes.
- Mesh data. This is usually produced by the pre-processor but may be modified by the solver program if some form of automatic mesh adaptivity has been carried out where extra nodes and elements have been added to the mesh or where nodes have been moved to try and improve the accuracy of the solution. It will include the coordinates of the nodes and the element connectivity list.
- Some form of ASCII file that reports on the progress of the solution. This file might include an echo of the input data from the pre-processor so that the input actually used by the solver can be checked, a repeat of the residual values, if there are any, and any user-programmed results, such as the displacements or stresses of a given node or the reaction forces. Accounting information such as the length of time that the solver took to run and the amount of disk resources used may also be listed.

Here, the concern is with how the actual structural data at all the nodes and elements can be analysed. Large quantities of this data are produced by the solver, especially if the mesh is complex and has a large number of nodes or elements, as is generally the case for an industrial problem. Only when small test cases are run is it possible to read the ASCII files that contain the solution and so for realistic problems computer graphics techniques are used to analyse the results visually.

6.8 USING COMPUTER GRAPHICS TO EVALUATE SOLUTIONS

6.8.1 Using Graphics Hardware

Before considering what can be done with computer graphics let us think about the hardware that is required to drive the software that generates the pictures as well as to display the pictures themselves. A typical hardware installation consists of the following devices:

- A screen or visual display unit (VDU) that is able to produce a grid of points in a variety of colours. These points are known as pixels (see Sec. 4.3.2). The resolution of the screen is determined by the number of pixels that can be displayed; most graphics screens can display a grid of 1000 pixels in the horizontal direction by 1000 pixels in the vertical direction. If the display is monochrome then each pixel can be shown as either black or white, whereas if the display is a colour device then each pixel can be displayed in one of several colours. Typically, 16 colours or even 256 colours are used. The screen could be part of a terminal attached to a computer or it could be part of a workstation.
- A keyboard that allows the user to interact with the software by typing commands and replying to questions from the software.
- A pointing device that enables a cursor to be moved around the screen. This pointing device may be a *mouse* which is a small device that senses movement either mechanically or optically, or it may be a simple set of four direction keys.
- A button box. This is used in the more expensive installations to manipulate the picture. The box has several knobs on it that can be used to rotate an existing picture about any of the three coordinate axes, or to zoom in and out or pan across the picture.

When the user runs the graphics software, the program activates the screen, keyboard, pointing device and button box in such a way that the user can develop an intuitive feel for the manipulation of the results.

6.8.2 Using Graphics Software

The graphics software itself is usually supplied as part of the structural analysis software package and is known as a post-processor. Sometimes, however, this software is combined together with the pre-processor to form a single interactive program that is used for both creating the computer model and post-processing. These programs enable a user to see the geometry of the structure, the mesh and the results of the simulation by producing pictures of the available data, usually in colour. Displaying the data in a visual way condenses the vast amount of information that a solver can generate into a usable format. As computer power becomes cheaper, graphics software is often run on interactive colour workstations that have sufficient display resolution for the task and also have enough of their own computer power to produce detailed pictures in a reasonable time without having an impact on other users on the network.

By entering commands the analyst can drive the post-processing software. These commands direct the software to build the required picture of the data on the graphics screen. Several commands may be needed to create a picture and, in many cases, the analyst will want to generate similar pictures from one analysis to the next. To prevent the user from re-entering a lengthy set of commands it is often possible for the software to read the commands from an ASCII file. This file can be created by the user with a text editor or it can be written by the software itself in some cases.

When generating the pictures, the stages that are followed are similar regardless of the type of data being displayed. The display process involves, first of all, displaying some part of the geometry or mesh on the screen. This may be a collection of the basic entities that comprise the geometrical hierarchy (see Fig. 5.5), or the boundaries of the mesh or even some part of the mesh itself, in either its original or its deformed state. Then, the picture is manipulated so that the required view is displayed before the solution itself is shown. This final display might be some of the stress data, shown as a set of vectors, or the contours of scalar variables such as the von Mises' stresses or even a magnified view of the displaced geometry. These three stages—show the geometry, modify the view and display the results—can be performed in any order but it is usual to display the actual results last of all. As this post-processing part of the analysis process is highly interactive, the user can often move between these three stages in a seemingly random fashion. However, for most simple cases, it is most useful if the order given above is followed. The following sections deal with each of these three stages in turn.

6.8.3 Plotting the Geometry

When the post-processing software is started it has to read the files of results and mesh data. Then the user has to find the required view. One way of doing this is to plot some part of the geometry, normally a part of the mesh used in the solution process, onto the graphics screen. This can be done by asking the program to display the basic entities used to create the mesh, if it has access to this data, or the boundaries of the mesh or the mesh itself. Exactly which of these is used depends on the capabilities of the software itself and the user's preference. A simple plot of the boundaries of the mesh is usually good enough at this stage.

However, a plot of the geometry or mesh can be used to check that the geometry looks like the physical situation and also to check the integrity of the mesh (see Sec. 5.13). 'Integrity' means that the mesh should both represent the required domain and be structured in the correct way. The display of the mesh shows a user the elements that have been used in the calculation procedure, and so any significant errors in the mesh or bad modelling practice can be found.

The way in which the mesh is displayed depends on the mesh structure that is being used. If the mesh has a regular structure then the local coordinate system and the node or element indices can be used to specify areas of the mesh just as was done in Sec. 6.2. Sheets of element faces can be defined in this way and then displayed. On the other hand, if an unstructured mesh is being used then the elements can be grouped in some way and the group projected onto some cutting plane in space. Another way of displaying the mesh is to draw only the free faces of the mesh.

6.8.4   Obtaining the Required View

Once the geometry has been plotted, the view of the geometry may well have to be manipulated. There are an infinite number of ways of looking at any image and so there must be some means of defining the exact view that is required. The picture on the screen is drawn as if a single eye is looking at the object being drawn. This situation leads the graphics software to require the user to define a few funda-mental pieces of data. This data can include such things as:

- the *target point*, which is the point in space at which the eye is looking
- the *eye position*, which is the point in space at which the viewing eye is placed
- the *up-direction*, which defines where the top of the picture should be
- the *viewing area*, which enables the apparent size of the objects in a view to be specified.

In Fig. 6.8, the target point is taken to be at the origin of a set of Cartesian axes. This target is shown being viewed by a single eye which can be placed in two different positions. Default values are always given by the software for the initial specification of both the target point and the eye position. These could be the origin and a point on the *x*-axis, such as eye position 1, respectively. When plotting data that relates to engineering work, the eye is normally at an infinite distance from the target and so the effects of perspective are not seen. This means that even though the eye position can be defined as a point in space the software actually places the eye at infinity on the same directional vector that joins the eye position and the target position. So, it can be seen that it is the combination of the eye position and the target point that defines the vector along which the eye looks. For some work, however, such as architectural drawing or aesthetic design, per-spective effects can be produced by the software and then the eye position is the actual point in space at which the eye is placed.

Defining these two positions in space is still not sufficient to specify the view of an object. Human beings have a sophisticated balance system and this gives us

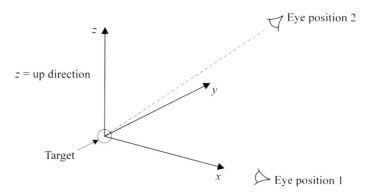

Figure 6.8 Target and eye positions.

information as to which is the vertical direction and so where up and down are. Computers are not as sophisticated and so they have to be told where the vertical direction is. This direction is also known as the up direction. In Fig. 6.8, the up direction is in the positive z-direction. Consequently, the post-processor must be told which direction the up direction is, if the pictures that it produces are to have the structure in a realistic orientation. One command usually enables this direction to be specified.

If the up direction cannot be specified to the post-processor, as is sometimes the case, then the picture has to be orientated by a series of rotations about the three coordinate axes. This is usually achieved by specifying the angles for each global coordinate axis, x, y and z, through which the axes are to be rotated. It is difficult to produce the correct view this way using a single command. Several attempts may be needed to get the picture right.

Once the eye position, target point and the picture orientation are known, the display software can take the three-dimensional data for the geometry or mesh and draw it on the screen, in what is, of course, a two-dimensional representation. This can be done in one of two ways. The original way was to transform the three-dimensional data into two-dimensional data using the post-processing software. This two-dimensional data can then be plotted. Many systems still use this technique, but a more recent way of handling the data is for the post-processing software to send the three-dimensional data to the display hardware, together with the current eye-position, target point and the vertical orientation. The transformation of the data from this set of three-dimensional vectors into a two-dimensional picture is then carried out within the hardware itself by a combination of both hardware and software, known as *firmware*. This local transformation is extremely fast as the firmware is dedicated to the task. Once the three-dimensional data is stored by the firmware it can be manipulated into further displays very easily and quickly, and this is where the button box, mentioned in Sec. 6.8.1, can be used very effectively to modify the target point, eye-position or orientation, signalling the firmware to produce the new pictures so fast that the objects can be moved in real time.

Quite often, attention must be focused on one particular area of the model, for example to see the detailed maximum principal stress pattern at a corner of an object. This can be done by changing the target position and the viewing area. The mechanics of doing this with the post-processor can vary, but there is nearly always a *zoom* command or a *centre* command. Figure 6.9 shows an example of the zoom command being used. This allows a rectangular window to be placed over the current view by defining the two ends of one of the diagonals of the window with the cursor. The software then modifies the target position and the view area to display the picture within the limits of the window. This is done while ensuring that the aspect ratio of the geometry is preserved. The centre command works in a similar way, but the user has to define the required centre of the new view, together with the magnification required, as shown in Fig. 6.10. By using these commands in the correct combination, the view of a mesh or the results can be infinitely varied. There are also ways of returning to the original view.

Original view   New view

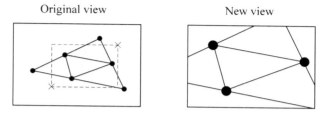

× Cursor-picked points

Figure 6.9 Changing the view by a zoom operation.

Original view   New view

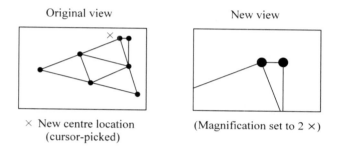

× New centre location   (Magnification set to 2 ×)
    (cursor-picked)

Figure 6.10 Changing the view by moving the centre of the display and applying a magnification factor.

When working with very complex meshes, and the associated results, the sheer volume of information displayed can be too great. The information content can be restricted by using the following techniques:

- *Volume clipping* This enables the user to give limits in the global coordinates x, y and z within which objects are displayed, but outside of which they are ignored.
- *Suppression of hidden lines* This calculates whether something that would be drawn is hidden from view by any other object, such as an element face. If the object is hidden from view it is not drawn. The displays that are generated using this method are often called hidden-line displays.

6.8.5   Displaying the Results

Having looked at how the geometry or mesh of the model can be displayed, and knowing how to orientate the view to give the desired picture, some of the results can be added to the picture. The results that can be viewed graphically have to be derived from the nodal displacement data.

With a mesh that has a regular structure the results data can be drawn for a sheet of elements or nodes, in the same way as the mesh can be drawn. Remember that this sheet may not be planar in global coordinate space, as even a mesh with a regular structure can be curved in space so that it fits around an object. When the mesh has an irregular structure the display of results is not so straightforward. As there is no simple way of referring to a group of elements, many post-processors allow the user to define a geometrical plane through the mesh onto which the results are interpolated. This plane is known as a *cutting plane*. Other ways of grouping elements can also be used, such as showing a hidden-line plot of the results which displays only those results on the boundaries of the mesh, or displaying the results for a restricted group of elements defined by creating a list of element numbers.

No matter which way is used to display the data, there are essentially three types of results display:

- *Vector plots*   show the vectors relating to the stress results.
- *Contour plots*   show contours of the scalar variables over the domain.
- *Deformed geometry plots*   show the deformed geometry in a form that magnifies the displacements.

The last case is simple, but the first two need some explanation. Note that the last case can be combined with either of the first two cases.

Dealing with vector plots first, the vectors are displayed within the picture as arrows in two dimensions. These plots are what are seen when the so-called wind arrows are shown on weather forecasts. Plotting stress information in this way can lead to confusing displays being produced as information is lost. The arrows that are drawn are the projections of a three-dimensional vector into two dimensions. Take, for example, a vector pointing directly out of the page; this would be displayed as a point. So that some of the lost information can be retrieved, the arrows are often colour coded to denote the absolute magnitude of the vector. Usually, red denotes a high stress and blue a low stress with intermediate shades denoting the stress values in between.

One other problem that has to be dealt with concerns the length of a typical vector arrow. Depending on the problem, the user will want the length of the arrows to give as informative a display as possible. This means that the user must scale the arrows appropriately, either by letting the computer draw some arrows and then scaling them, or by giving the computer a typical stress which might represent, say, 10 per cent of the screen width.

For meshes that have very dense element distributions the arrows may be so close together that too much is displayed and the useful information is obliterated. This can be overcome by the software interpolating the stress data onto a coarse, regular grid of points. The user specifies the distance between the points in the grid, and the arrows are drawn at the points. One problem with this type of display is that the true nature of the computed stress field can be hidden from the user. Sometimes it is better to display the data at the positions at which it was calculated. (In later chapters vector plots have not been used for the reasons above as well as the fact that monochrome vector plots are extremely difficult to interpret.)

Contour plots are pictures of the lines of constant scalar value of some variable plotted through the domain. They are similar to the isobars that are seen on maps for weather forecasts. Little interaction is required to produce these plots, except perhaps to specify the number of contours that are to be drawn and their values, which need not be specified with a constant increment. Typically, about 10 contours are calculated, and again these are colour coded in the picture to show the value of the variable on the contour. A coding scheme similar to that used for the magnitude of a vector is used in this case as well. Sometimes, the contour levels can be chosen by the user to give the required values. This is done where several separate pictures of contours have to be produced to create the required display, and it provides a consistent display.

A variation of the contour plot is to use a surface plot. This is generated by displaying a three-dimensional surface, the height of which above a plane is a measure of some variable. This variable should be a function of the two dimensions that describe the plane. In effect, the display shows a series of mountains and valleys.

6.9 CHECKING A SOLUTION FOR ACCURACY

When analysing the results of a simulation, certain pieces of information are required. For example, a prediction of the von Mises' equivalent stress at some point in the structure may be needed.

As a check on the quality of the result:

- The stress field and displacements should look qualitatively correct. If these simple checks show that there might be problems with the quality of the results then users should consider checking their input data and changing their models, if necessary, before re-running the solver program.

6.10 MAKING REFINEMENTS

If it looks likely that a model must be refined, a user must consider the advantages of producing a better prediction against the cost constraint of repeating the whole simulation process. Quite often even crude models can give large amounts of new and useful information to a user. This might prove adequate for the purposes of some users but not for others. It all depends on the application under consideration.

The process of refining a model can include increasing the density of nodes in a given area so that the changes of the displacement and hence stress in that area can be more accurately captured, for example, in the region of a hole or other structural discontinuity. In terms of effort, this involves a large amount of work, as it involves rebuilding the mesh of the domain, either by repeating one of the mesh generation processes described in Chapter 5, or by using an adaptive meshing process. Once the mesh is built the problem specification within the pre-processor and the setting of the boundary conditions has to be carried out again, transforming the data generated as part of the original specification process onto the new mesh.

A systematic way of increasing the mesh density for a mesh with a regular structure is to double the number of elements in each of the local mesh directions. Similar refinement schemes can also be carried out with unstructured meshes, for example, by placing a new node at the centroid of each element and then remeshing. With the new mesh a solution is calculated, and the results obtained. When the results do not vary in global terms from one mesh refinement to the next then the results are said to be mesh independent. While it is always desirable for the results to be independent of the mesh size, for many industrial problems this is not always possible as the constraints in terms of cost or time or computer capacity are too great.

6.11 FAILURE CRITERIA

Stress results from an analysis can be used to determine whether or not the material has failed. This means that, for a successful design, the geometry of the structure has to be modified to relieve damaging stress concentrations. This failure analysis is usually achieved by transforming the nodal stresses, given with respect to the global coordinate directions, into the principal stresses and substituting these stress values into a failure criterion appropriate to the material in the structure.

The most common of the failure criteria found in packages is that due to von Mises. It is a criterion relevant to ductile materials, such as metal alloys, that fail by the single mechanism of a material yielding. Output from the criterion is in the form of a stress, known as the von Mises' stress, which, if it exceeds an allowable material stress (often the tensile yield strength), tells the analyst that the material has failed.

For further details on the formulation and application of failure criteria for structural materials consult texts specific to the material in question. Often the software manuals provide an introduction to the failure criteria available to the analyst, but be aware that this information does not generally go into sufficient depth to mention any pitfalls that might occur when these criteria are used such that the limitations of the criteria are ignored.

7

7

SOME EXAMPLE PROBLEMS

This chapter considers example problems that illustrate the techniques discussed
in the previous chapters with regard to small displacement linear elastic analysis.
Simple examples will be discussed before moving to the more complex examples
of Chapter 8 and the industrial examples of Chapter 9. The effects of dead weight
and real imperfections have not been included in the examples.

7.1 THE BRAZILIAN OR SPLIT TENSILE TEST

This problem is contrived to be two-dimensional and of plane stress type. It
illustrates a simple use of the finite element method with an isotropic homogeneous material. Further, the equations developed in Chapter 2 yield a classical
elasticity solution against which the finite element solution can be compared.
There is also a physical, practical use to this test as it is a standard test method
for the tensile strength of brittle materials.

Through this example it is hoped to demonstrate the use of symmetry, the
application of a concentrated point load which is a stress singularity, the mesh
specification for an unstructured, or free, mesh using triangular elements, the
modelling of curved boundary by quadratic elements, the grading of a mesh for
variable stress gradients and convergence by p-refinement methods.

Over the years a considerable amount of experience has been gained by the
authors by using this example as the subject of undergraduate assignments. It is
particularly suitable for this as any finite element solution can be compared with
both the classical solution and the behaviour of specimens in physical tests. From
this comparison the success of finite elements in modelling the problem is easily
demonstrated.

7.1.1 Problem Definition

Figure 7.1 shows both the geometry and the loading boundary conditions for the Brazilian test. A disc of material is compressed by diametrically opposite concentrated (or point) loads, where d is the diameter of the disc and P is the vertical load acting along the line AOB. Here, it is assumed that the diameter d is very much greater than the thickness of the disc t such that the problem can be reduced to a two-dimensional case with plane stress. Furthermore, it is assumed that the values of P, d and t are such that there is no nonlinearity induced in the material and so the problem can be modelled as being one with linear elasticity and small displacements. Hence a linear static analysis can be performed.

Although the presence of circular symmetry suggests that the use of a polar coordinate system may be convenient, the results presented here are defined with reference to a Cartesian coordinate system placed at the origin O. Note that the planes of mirror symmetry for this loading case are labelled AOB and COD.

In practice, the Brazilian test method is used to determine the tensile strength of brittle materials such as ceramics and concretes, because a favourable stress distribution is generated in the specimen. In such a test, a disc (or cylinder) is compressed between two hard parallel metal platens, using a standard testing machine, until ultimate failure occurs and the specimen splits along diameter AOB.

This example is suitable as an undergraduate assignment where plots (in a suitable non-dimensional form) of the stresses σ_x, σ_y, τ_{xy} along the diameters AOB and COD have to be created using both a finite element analysis and a classical elasticity solution.

7.1.2 Classical Elasticity Solution

By following the Airy stress function approach outlined in Sec. 2.3.2, a classical elasticity solution for this problem can be developed provided that the material is

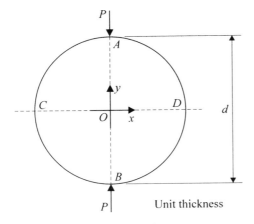

Figure 7.1 The arrangement for the Brazilian tensile test.

homogeneous and isotropic (Timoshenko and Goodier, 1988). It is shown that the stresses (σ_x, σ_y, τ_{xy}) are as follows:

Plane *COD*

$$\sigma_{y(CD)} \bigg/ \left(\frac{2P}{\pi d}\right) = 1 - \frac{4d^4}{(d^2 + 4x^2)^2}$$

$$\sigma_{x(CD)} \bigg/ \left(\frac{2P}{\pi d}\right) = 1 - \frac{16x^2 d^2}{(d^2 + 4x^2)^2} \tag{7.1}$$

$$\tau_{xy(CD)} \bigg/ \left(\frac{2P}{\pi d}\right) = 0$$

Plane *AOB*

$$\sigma_{y(AB)} \bigg/ \left(\frac{2P}{\pi d}\right) = 1 - \frac{d}{d/2 + y} - \frac{d}{d/2 - y}$$

$$\sigma_{x(AB)} \bigg/ \left(\frac{2P}{\pi d}\right) = 1 \tag{7.2}$$

$$\tau_{xy(AB)} \bigg/ \left(\frac{2P}{\pi d}\right) = 0$$

Note that $\sigma_{x(AB)}$ is a *constant tensile stress* , and that it is this feature of the stress distribution that makes the Brazilian test so useful in measuring the tensile strength of brittle materials. Also note that the thickness t is taken to be unity and so does not appear in the expressions. Finally, it is convenient to use the nondimensionalized forms given in (7.1) and (7.2) when plotting stress results as these forms are independent of the values of load P and diameter d.

From (7.1) and (7.2) several interesting features of the solution can be seen:

- σ_x is tensile and constant along diameter *AOB*.
- τ_{xy} is zero along diameters *AOB* and *COD*.
- Both σ_x and σ_y vary gradually along *COD*. These stresses are also seen to be zero at points C and D which are located at stress-free surfaces.
- The value of σ_y tends to minus infinity as the point of loading is approached. This represents a mathematical singularity resulting from the assumption that all the load is applied at a point. Physically this is unrealistic as the load cannot be applied over an infinitesimally small area. This analysis does not account for any yielding of the material, such as would be included to ensure that the area of the loading was acceptable if the material were a ductile metal. In reality, where the infinite stress occurs, local deformation or failure increases the area through which the load P is transmitted, as shown in Fig. 7.2. This results in the realistic situation of a distributed pressure loading being applied over a small finite area.
- The solutions along diameters *AOB* and *COD* are symmetric about point O.

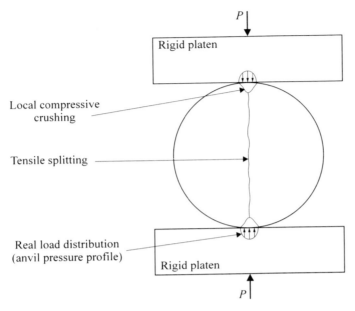

Figure 7.2 Failure of the disc.

7.1.3 Definition of the Finite Element Model

Having looked at some of the background to the Brazilian test and having discussed the classical solution to this problem, the finite element modelling process described in overview in Sec. 4.1 can be begun. The first stage in this process is the initial thinking stage where the structural problem is described (see Sec. 2.4) in terms of the analysis requirements, the structural geometry, the material properties and the loads and restraints on the structure.

As a classical solution to this problem in the form of (7.1) and (7.2) for the stresses along two diameters exists, it is appropriate to build a finite element model and compare the stresses generated by this model with those of the classical solution. More complicated requirements for other problems will be described later.

Turning to the geometry of the problem, Fig. 7.1 shows the actual Brazilian test, but as the shear stresses are zero along diameters AOB and COD, engineering judgement can be used to determine that there is quarter-plane symmetry about these diameters. Hence, only one-quarter of the geometry needs to be considered. For convenience the positive x–y quadrant is chosen as shown in Fig. 7.3. Finally, as a plane stress problem is required, the thickness of the disc in the third dimension is taken to be unity.

This test is used on a wide variety of materials and, looking at (7.1) and (7.2), it can be seen that the stresses should be and are, for the reasons discussed in

Chapter 2, independent of the value of the material elastic constants. This means that it does not matter which material properties are chosen so long as they are isotropic.

As to the physical loads and restraints on the geometry, Fig. 7.3 shows the situation for the quarter disc. If the full load to the disc is P then the model requires a load of $P/2$ to be applied. On loading the disc, the point at the origin O can have neither u- nor v-displacement, whereas any other point along the vertical line OA is free to move in the y-direction but not in the x-direction where symmetry forces the disc to be restrained. Similarly, along line OD any point except the origin is free to move in the x-direction but not in the y-direction. Hence along OA the v-displacement is free and the u-displacement is zero and, along OD the u-displacement is free and the v-displacement is zero.

Having thought about the physical problem, the computer modelling can begin. From Sec. 4.1 the second stage of the analysis process is generating a mesh, although as was seen in Chapter 5 this includes developing a model of the geometry of the structure. In this case the geometry is very simple and can be modelled by two straight lines and a circular arc. Usually, defining the endpoints of the lines and three points associated with the arc, say the centre, start and endpoints is all that is required. To produce the mesh the volume or, in this case, area of the structure must be defined. This is done by forming a closed loop from the two lines and the arc in sequence. For this problem the origin of the Cartesian coordinate system has been taken as point O and the radius of the disc as 500 mm.

Once the geometry has been defined a mesh must be created, but this mesh must be suitable for the problem being modelled. The following are pointers as to what constitutes a suitable mesh:

- The stress gradients (and strain gradients) are going to be very high in the region close to the point of loading and so the smallest elements in the mesh are

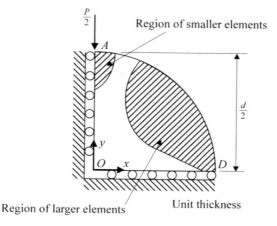

Figure 7.3 Boundary conditions for the Brazilian test.

going to be in this region. Elsewhere the gradients are relatively low and so the element size can be larger. Further, as nodal stress values along the boundaries of the model are to be compared during the results analysis there must be an adequate number of nodes (or elements) along the radii OA and OD. Finally, the curved boundary, except at the load point, is stress free, having zero traction. This means that the largest elements can be placed along this boundary. All of this means that a graded mesh must be used with the smallest elements near the point of loading and the largest elements along the curved boundary, as shown in Fig. 7.3.

- The shape of the geometry, which is topologically triangular, suggests that it might be sensible to use triangular elements. Also, as one of the sides is curved these elements should be of at least quadratic order.

These points indicate that an unstructured mesh is the most suitable, with some form a grading applied to the element sizes. In Sec. 5.9.1 it was stated that, to prevent numerical errors creeping into the solution, it is advisable not to let the volume of the elements in a mesh differ by more than 30 times. To enforce this the typical element dimensions have been specified to be $\frac{1}{20}$ of the disc radius near the load point A and $\frac{1}{4}$ of the radius near the curved boundary. As all of the elements are approximately equilateral triangles, and are not too distorted from their master shape (another desirable feature when modelling), this gives a maximum volume difference of 25 times.

Use of an automatic free mesh generator with these element dimensions and the element type set to quadratic plane stress triangular elements produces the mesh shown in Fig. 7.4. In this mesh there are 95 elements and 220 nodes. Based on our experience of marking undergraduate assignments, good finite element results can be obtained with a 'well-constructed' mesh of around 100 elements, and so the mesh produced here should be suitable. Note, however, that while quadratic elements have the ability to fit quadratic curves, it appears from Fig. 7.4 that they are unable to model exactly the circular boundary here.

It is always good practice to solve a problem using two different finite element models as this gives some idea of the accuracy of a solution. In this case, by changing the order of the element from quadratic to linear, a second mesh is obtained that can be analysed. This is an example of the process of p-refinement, which should show that a solution is converging as the order of the element increases.

Once the mesh is built the next stage can be carried out where the numerical problem is defined on the mesh. This involves:

- Defining the load to act at the correct node, to be of magnitude $P/2$ and to be negative, to signal to the software that it acts in the negative y-direction. In this case the load at the node was set to be -0.785398 N (i.e $-\pi/4$).
- Assigning all the nodes along the quarter planes of symmetry to have the relevant displacement boundary conditions as already discussed. For each node one of its displacements is fixed and is therefore assigned a zero displacement while the other displacement is set to be free.

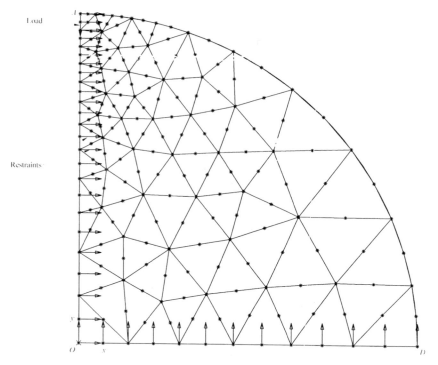

Figure 7.4 A finite element mesh with 95 quadratic plane stress triangular elements.

- Choosing default material properties of structural steel.
- Setting default physical properties with an element thickness of 1 mm.

Now the solver can be run and the results obtained. For this small model, computer time is not an issue and provided that the model is suitably restrained the solution should be generated with little trouble.

7.1.4 Finite Element Results

There are several graphical forms of output from commercial finite element software: nodal displacements, nodal stresses, and contour and vector plots of stresses. Here only the first two of these will be dealt with, as contour plots will be discussed when presenting solutions for the next example, the pressure vessel.

Figure 7.5 shows the deformed shape of the disc superimposed on the undeformed shape as calculated by the model with quadratic elements. It is clear that the deformation, and its rate of rate change with distance, is concentrated in a region below the point of loading. This provides even more evidence for the necessity of having a graded mesh. In this figure, the deformed shape is made clearly visible by multiplying the deformation by a factor of many

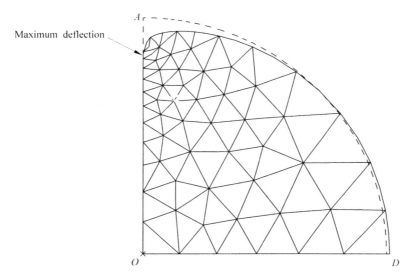

Figure 7.5 The deformed shape predicted by analysis of the quadratic element mesh.

thousands. Those not familiar with finite element analysis might mistakenly use this figure to convince themselves that the material has gross plasticity, and so it is worth emphasizing here that in a linear static analysis the software has no understanding of the strength of a material when solving (1.2). In other words, the stress results is always based on the appropriate stress–strain relationship, which for plane stress is (2.29). It is only when the modeller instructs the software to create a failure criterion contour plot, for example with a ductile metal, von Mises' stress contours might be used, that some idea as to which areas have yielded is obtained.

Profiles of finite element stress values along OA and OD are plotted in Figs 7.6–7.12, together with the classical values from (7.1) and (7.2). These stresses are shown as nodal values of σ_x, σ_y and τ_{xy} with a linear interpolation used to find the stresses at the internodal positions. This has been done for both the linear and the quadratic models. Sec. 3.7.8 stated that element stresses are calculated most accurately at the Gauss points. To obtain the nodal stresses shown, element stresses at the nodal positions have first been found from the Gauss point stresses. Hence, where a node is connected to two or more elements there are two or more stress values. To resolve this into a set of single values at each node, some form of weighting must be performed. The simplest approach is to average the element values where necessary, but this does not take account of any size disparity between connecting elements. Consequently, the software might weight the nodal values by element volume or area, which is more accurate. This aspect of nodal stress calculation can have a direct bearing on the accuracy obtained.

It is apparent from the solutions shown in Figs 7.6–7.12 that the results from the quadratic element model are superior to those from the linear element model.

There are two reasons for this. First, a quadratic element allows a linear strain and stress variation within it, while a linear element allows only constant values of strain and stress. Problems with rapidly changing strain and stress gradients are, therefore, modelled with better accuracy when quadratic elements are used. Second, as more degrees of freedom are calculated within a discrete finite element model so a better approximation is made to the actual problem, which is continuous with an infinite number of degrees of freedom. For the mesh generated here there are 95 elements with 402 active degrees of freedom in the quadratic element model but only 106 active degrees of freedom in the linear element model. If this test case is to provide a direct comparison between the performance of these two element types, then the effects of not only a p-refinement but also of an h-refinement, where the number of elements within the mesh is systematically increased, need to be compared.

Having seen that by changing the order of the element from linear to quadratic the number of variables calculated is increased by a factor of four, it may be expected that the solution time will increase substantially. In this case the actual increase in computer effort is only 25 per cent. Such an increase is not a modelling issue as the computer effort involved is so low in both cases. However, when a model requires thousands of elements and, hence, thousands of variables, such increases may be extremely significant, limiting the extent of the refinement of the mesh. In such cases analysts are advised to use fewer quadratic elements rather than a large number of linear elements.

7.1.5 Comparison of Classical and Finite Element Solutions

Now the finite element solution can be compared to the classical solution. To do this results along OA will be considered and then those along OD.

Along radii OA (y-axis) Figures 7.6–7.9 show the stresses along OA for both linear and quadratic finite element models and for the classical solution. The stress σ_x is illustrated in Fig. 7.6, with a magnified form of this being shown in Fig. 7.7. From point O for some 80 per cent of the radius the comparison between all three solutions is very good, although Fig. 7.7 shows that the linear element solution is slightly low compared to the classical solution. However, moving closer to the load point A the linear element solution departs from the classical solution quite quickly, while the quadratic element solution starts to oscillate in an attempt to stay close to the classical solution before falling rapidly negative to the physical solution at A. This difference in form of the σ_x profiles generated using linear elements, giving a smooth form, and quadratic elements, giving an oscillatory form, is typical of the behaviour for these element types in this situation and is also independent of the form of the mesh. Note that when the number of elements in a model is increased, the location along the radius where the stresses become negative moves closer to point A, but never reaches it.

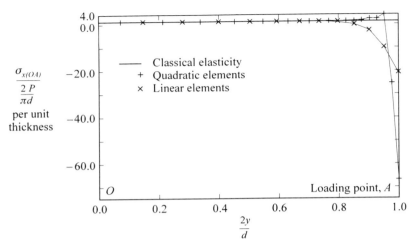

Figure 7.6 The nondimensional direct stress in the x-direction along plane OA.

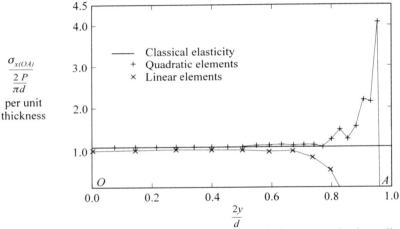

Figure 7.7 A magnified view of the nondimensional direct stress in the x-direction along OA.

Figure 7.8 shows the comparison for σ_y. The quadratic solution is seen to be quite accurate even in the region of the load point A. However, the comparison for τ_{xy} in Fig. 7.9 shows similar trends to those for σ_x.

Along radii OD (x-axis) Figures 7.10–7.12 show the stress comparisons along OD. These figures show that there is a small difference between the classical and finite element solutions when the elements are quadratic. This excellent correlation shows that the mesh construction with 6 elements, i.e. 13 degrees of freedom,

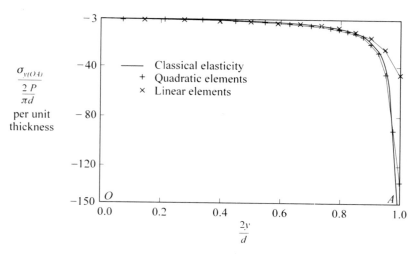

Figure 7.8 The nondimensional direct stress in the y-direction along plane OA.

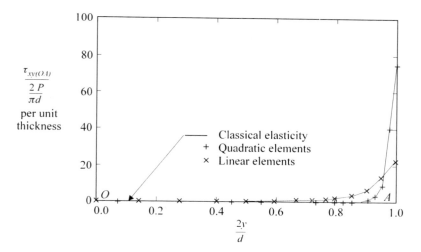

Figure 7.9 The nondimensional shear stress in the xy-plane along plane OA.

along OD is acceptable and does not require further optimization. The form of the stress profiles is similar when the element type is linear, but there is a significant error in the values of stresses σ_x and τ_{xy}. Improvement in the comparison may be obtained by increasing the number of elements in the linear element model from the present 6 elements, i.e. 7 degrees of freedom.

Clearly, the largest differences between the solutions on both axes occur along OA in the region near the load point A. To explain the behaviour of the finite element models near this point, the analyst must consider what the elements are

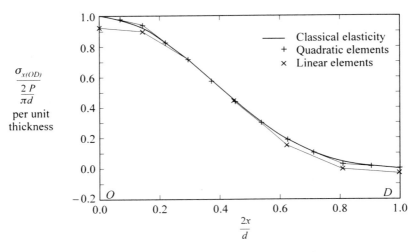

Figure 7.10 The nondimensional direct stress in the x-direction along plane OD.

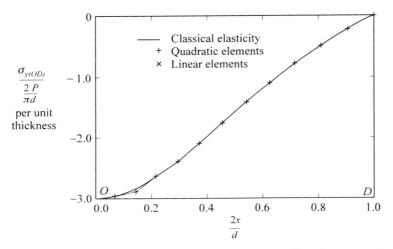

Figure 7.11 The nondimensional direct stress in the y-direction along plane OD.

actually doing. Adjacent to A the elements have the effect of averaging out the very high strain and stress gradients that occur in this region. This leads to considerable inaccuracy, so it is not too surprising to find a large difference between classical and finite element solutions in this region. The classical solution satisfies the mathematical modelling assumptions, equilibrium equations, compatibility equations and stress–strain relationships discussed in Chapter 2, and so it is exact at every point within the continuum—at least this is so for the perfect case being considered here.

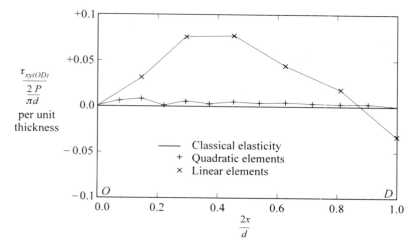

Figure 7.12 The nondimensional shear stress in the xy-plane along plane OD.

Returning to the comparison of the finite element solution with the classical solution, it is often thought that the finite element results must be wrong as they do not match the classical solution everywhere. This diagnosis is false in an engineering context, and hinges on the meaning of the word 'exact' when used in the context of a classical elasticity solution. It is a shock for many people when they are shown that the finite element solution gives a better picture of the physical reality in a practical test. There, it is found that the splitting along diameter AOB occurs only after local crushing under the loading platens. This is illustrated in Fig. 7.2 where the failure mechanism for a specimen is shown.

Thus, it is found that the finite element solution, although very similar to the classical at most points along the quarter symmetry planes, does provide valuable information as to what is happening in a real test under the point of loading.

7.1.6   Possible Improvements to the Finite Element Modelling

There are three modifications that may be made to this model to improve the accuracy of the solution:

- Using more quadratic elements in the region near to the load point.
- Rearranging of the mesh in the region adjacent to the load point such that instead of the loaded node being connected to a single element, as shown in Fig. 7.4, there are two or more elements connected to this node, as shown in Fig. 7.13.
- Distributing the load P as a surface pressure having the profile shown in Fig. 7.2. This gives a more realistic load case, but for such a surface stress distribution to be present the geometry of the disc must become flatter over the area of contact.

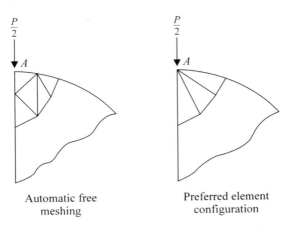

Automatic free
meshing

Preferred element
configuration

Figure 7.13 Mesh construction at the loading point.

Knight (1993) shows a variety of simple models using such techniques. Note, however, that in typical situations a concentrated force is applied at a single node. This is a common simplification for a surface pressure distributed over a small area, and if the actual surface area is smaller than the face area of elements attached to the node, then this is acceptable. However, if this causes a stress concentration that is too high, then the forces must be distributed over two or more nodes placed in such a way that the boundary stress is limited to the value of expected surface pressure.

7.2   A PRESSURE VESSEL

In this example, the analysis of a pressure vessel—an aerosol bottle—is considered. This is a real design problem that illustrates the use of general shell elements to model a bottle with a novel elliptic cross-section which has curved geometry and variable wall thickness. The bottle is made of a polymer material that is isotropic and homogeneous. In terms of modelling, the example includes the use of mapped meshing with quadrilateral elements, mirror symmetry, pressure loading and h-refinement. Difficulties occur as there are discontinuities in the geometry that give rise to poor local modelling in these regions. In particular an analysis will be made to produce a design with a suitable creep performance.

7.2.1   Problem Definition

As before, attempts must be made to build a specification for the analysis in the way that was discussed in Sec. 2.4. This means that the requirements of the analysis, the geometry of the structure, the materials and the physical loads and restraints on the system must be defined. Again it must be emphasized that this stage of the analysis process is done before any computing takes place.

Requirements   It is proposed to manufacture a domestic aerosol spray, the main container or vessel of which is to be made from a thermoplastic polymer material. At the initial design stage a finite element analysis is required to ensure that the vessel meets certain design parameters in a global sense. As will be seen from the geometry, there are discontinuities in the vessel design but for this global analysis the effects of these will be ignored to some extent. This initial analysis should show the following when the vessel body is pressurized:

- The maximum tensile stress in the vessel body, i.e. not in the complex area of the head and shoulder of the vessel, will be less than $20\,\mathrm{N\,mm^{-2}}$ (MPa). If this is achieved then the creep behaviour of the polymer can be neglected and so its long-term behaviour is not a consideration. If this is the case then the materials can be assumed to have linear elastic properties, making this simple finite element analysis valid.
- The maximum distortion from the unpressurized shape at any cross-section will be not more than 1 mm, i.e. the bottle when measured by a calliper gauge will be no more than 1 mm larger across the major or minor axis when pressurized.

Geometry   The thin-walled vessel, as illustrated in Fig. 7.14, consists of two injection-moulded parts welded together. These two parts are the body and the head–shoulder combination. At the top of the head section there is a flat steel cap. Nominal mid-surface dimensions and the average wall thicknesses, labelled t, are shown in Fig. 7.14. Note however, that the cross-section of the body and the shoulder of the vessel at the body–shoulder junction is elliptical with a major-to-minor axis ratio of 1.2. The shoulder cross-section undergoes transition from an elliptical section to a circular one at the shoulder–head junction, and the head is of a circular cross-section throughout. Note also that the base of the body is domed such that the dome protrudes into the volume of the vessel.

At the junctions between the dome and side walls, body and shoulder, shoulder and head and head and cap local effects occur, but these are not of concern here. These local effects are evaluated in detail using further finite element analyses after the global performance has been assured with the current analysis.

Materials   The thermoplastic polymer material can be assumed to have isotropic properties with a Young's modulus of $2.5\,\mathrm{RN\,mm^{-2}}$ (GPa) and a Poisson's ratio of 0.25. It is clear that engineering experience and judgement is often needed. The material under test (coupon) is likely to have a different modulus, although this is not guaranteed, to that of the material of the vessel, and this difference could be the cause of different processing conditions. The modulus in the test is not necessarily higher, but as the FE results are used in the design, the deformation should be overestimated if a safe design is to be

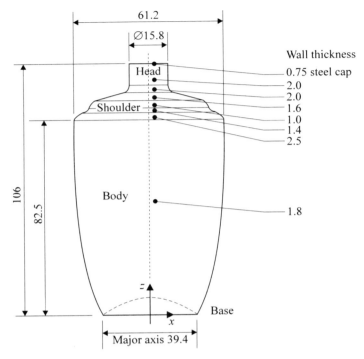

Figure 7.14 Geometry of the pressure vessel. (Dimensions in millimetres.)

ensured. There is an additional benefit in that the inherently over-stiff model that results from using finite elements formulated using assumed displacement fields, as noted in Sec. 3.5.1, is partly compensated for.

For the steel cap the standard properties for steel, i.e. a Young's modulus of $207 \, \text{kN} \, \text{mm}^{-2}$ and a Poisson's ratio of 0.29, can be assumed. The importance in any finite element analysis of choosing relevant and reliable material properties cannot be overstated and this issue is discussed in Sec. 2.2.6. If an analyst has a concern about any of the properties (strengths or moduli) it would always be sensible to choose values that are known to be underestimates. To choose a strength that is higher than the material can possess is to introduce a modelling error whose presence may well not be observed until the engineering product has failed.

Loads and restraints When the vessel is charged with its gas–liquid mixture, the maximum design pressure is to $0.5 \, \text{N} \, \text{mm}^{-2}$ ($5 \times 10^5 \, \text{Pa}$). This pressure acts equally on all of the surfaces of the vessel. As the cross-section of the vessel is elliptical or circular, a full three-dimensional model is not required if symmetry is taken into account. If the cross-section were circular throughout then an axisymmetric model would be suitable but the elliptical cross-section leads to mirror symmetry along only the planes of the major and minor axes. Hence quarter-symmetry can be used to allow modelling with only one quarter of the vessel. The actual boundary conditions required to achieve this will be discussed later.

7.2.2 Geometry and Mesh Definition

Before considering the creation of the computer model of the geometry and the mesh itself, both of which are fairly straightforward here, it is useful to think about the choice of element. In this case the geometry is curved and the wall thickness is very much less than the overall dimensions of the vessel. A reasonable fit to the curved geometry is achieved only if the order of the elements is quadratic or higher. Equally, it is acceptable here to use thin-shell elements, as the material is sufficiently thin and the effects of combined bending and membrane action on the displacement must be captured.

Both quadrilateral and triangular elements have been designed to give approximate solutions to Mindlin shell theory (Mindlin, 1951), where a quadratic distribution of the transverse (through-thickness) shear stress resultants is included. For curved shells, the Ahmad formulation (Ahmad, Irons and Zienkiewicz, 1970), is employed using standard eight-noded shape functions. These elements pass the patch tests for both membrane and bending actions. Note that each node has six degrees of freedom, with the displacements u, v and w being in the directions of the nodal Cartesian coordinate system axes and having dimensions of length, and the rotations θ_x, θ_y and θ_z being about the nodal coordinate system axes and measured in radians.

In Sec. 3.8.2 it was noted that there are many element formulations for general thin-shell problems, none of which exactly satisfies the requirements listed in Sec. 3.5.2 for an element to model faithfully the deformation of a continuum. As a result of this, the solutions presented here with the elements described in the previous paragraph may be quite different from those determined with another thin-shell element.

Returning to the actual geometry being considered, the data in Fig. 7.14 has been obtained from a set of engineering drawings. Note the choice of axis system where the z-axis is parallel to the centroidal axis of the vessel, the x-axis is in the direction of the major axis of the elliptical cross-section and the y-axis is in the direction of the minor axis.

As has been already mentioned, the model exhibits quarter-symmetry for the loading assumed here and so only one-quarter of the vessel need be modelled. In Fig. 7.15, the model is shown in the positive quadrant of the Cartesian system with the origin at the centre of the vessel's base.

To build the geometry the hierarchy of point to edge to surface has been used. This is done by determining key points from the engineering drawings and creating curves in the form of arcs and splines such that the 10 surfaces shown in Fig. 7.15 can each be defined from the bounding loop of curves. Note that nine of the surfaces are topologically quadrilateral and so they are suitable for the building of a mapped mesh of quadrilateral elements.

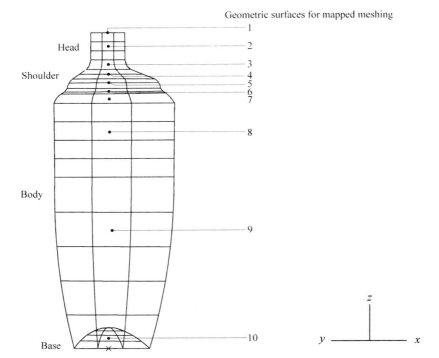

Figure 7.15 A coarse mesh of the pressure vessel.

Quadratic, quadrilateral elements, and their associated nodes, can now be placed on these surfaces, by defining the numbers of elements along each edge of a surface and the element thickness. Note that the elements within each surface have a constant thickness which is the average value of thickness for the region being considered (see Fig. 7.14). When all the surfaces have been meshed with quadrilaterals, triangular quadratic elements are created to complete the mesh where the dome of the base meets the z-axis and to model the steel cap.

A coarse mesh is shown in its complete form in Fig. 7.15 where there are three elements around the quadrant in the circumferential direction and so each element represents a surface which subtends an angle of $30°$. The number of elements through the height, i.e. in the z-direction, was chosen to make the overall mesh construction acceptable. For the distribution shown, it can be seen that the distortion of the elements in the main body is not large, thereby minimizing the errors in the computation resulting from distortion of the elements from their parent shape. Note, however, that the element distortions are larger in the shoulder and head regions of the vessel. In these regions the effect of element distortion is not too important as the stress results in these regions will not be accurate because of the continually changing wall thickness.

This coarse mesh has 78 elements (72 quadrilateral and 6 triangular), 281 nodes and 1086 degrees of freedom. Given the coarse distribution of elements some geometrical errors might be expected in the calculation. Figure 7.16 shows a section of the mesh in the transition region between the head and shoulder. The curves defining the geometry of the surfaces and the edge of the mesh can be seen to be different. Again it is clear that while quadratic elements have curved sides, they cannot, in general, fit a curved geometry exactly. The difference, however, may well be small enough to be ignored.

As a check on the convergence of the analysis, a second mesh has been constructed and this is shown in Fig. 7.17. Exactly the same methodology has been used to create this mesh but double the number of elements have been placed along the edges of each surface, and so the element density must be doubled. As this finer mesh is, effectively, the original mesh with each element split into four elements, the process of mesh refinement is known as h-refinement.

The finer mesh has 300 elements (288 quadrilateral and 12 triangular), 989 nodes and 5934 degrees of freedom.

7.2.3 Finishing the Finite Element Model

Once the mesh has been built appropriate material properties must be defined, if they have not already been defined as part of the mesh creation process. Then the correct boundary conditions are created for the model. Note that for both meshes shown in Figs. 7.15 and 7.17 the physical boundary conditions are the same, even though they are applied to different nodes and elements.

First, the internal pressure within the vessel at the design condition must be represented. In general, commercial software packages have the option to input a

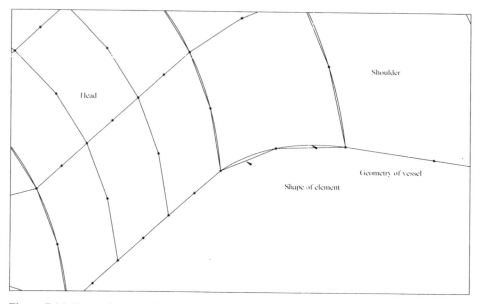

Figure 7.16 Errors in modelling the geometry of the pressure vessel.

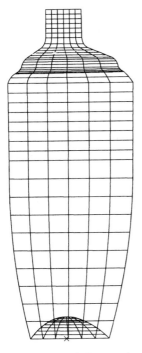

Figure 7.17 A fine mesh of the pressure vessel.

pressure loading directly. This is usually done by assigning a single value of pressure, in this case $0.5 \, \text{N} \, \text{mm}^{-2}$, to a series of element faces. The sign of the pressure loading defines the direction in which the pressure acts, with a positive sign denoting that the pressure acts in the directions of the positive axes of the Cartesian coordinate system. This pressure loading over the surface of an element is then converted by the software into a consistent set of nodal forces. Some loss of accuracy is introduced at this stage when the elements have curved sides and surfaces, as was discussed in Sec. 3.7.7.

To apply the pressure boundary condition all the elements are formed into a group, and for all members of the group the pressure value is set. Figure 7.18 shows the element pressure vectors on the coarse mesh, with all the vectors pointing out of the vessel, thereby indicating that the pressure within the vessel is greater than atmospheric pressure.

Displacement boundary conditions also need to be specified, to prevent rigid body motion and to enforce the correct deformation along the planes of symmetry. For a full three-dimensional model of the vessel, if the modelling of the pressure loading is exact, then the net force in all directions is zero. Equally, in this quarter-symmetric case, the overall force in the z-direction should be zero. Unfortunately, errors occur in the way in which the pressure loading is represented in the finite element analysis and, even though they are small, these errors

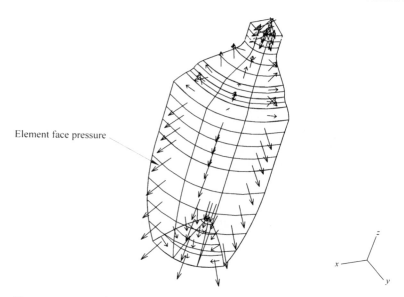

Figure 7.18 Pressure loading on the vessel.

ensure that the net integrated force due to the pressure loading it is not zero. Therefore, to restrain the vessel from moving bodily in the z-direction, at the nodes where the dome meets the body of the vessel, the w-displacement has been set to zero. This can be seen in Fig. 7.19 where a single arrowhead from a node denotes a fixed displacement.

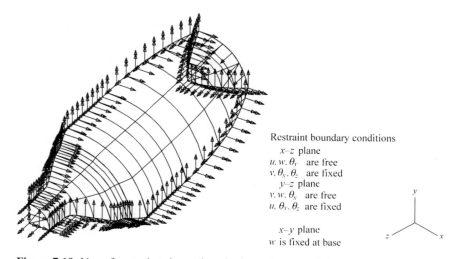

Restraint boundary conditions
  x–z plane
u, w, θ_v are free
v, θ_x, θ_z are fixed
  y–z plane
v, w, θ_x are free
u, θ_y, θ_z are fixed

  x–y plane
w is fixed at base

Figure 7.19 Use of restraints in setting the boundary conditions on the vessel.

Turning to the symmetry planes, let us consider a node lying along the edge of the vessel in the x–z plane. For symmetry to be enforced, these nodes must be free to move only in the x–z plane. This is done by fixing the v-displacement to be zero, shown in Fig. 7.19 by a single arrowhead again, and allowing the u- and w-displacements to be free. Also the rotation θ_y must be free and the rotations θ_x and θ_z must be fixed to be zero. These rotational boundary conditions are analogous to zero slope at the point of loading for a simply supported beam with a transverse point load. Using similar arguments for the y–z plane, the displacement boundary conditions at each node in this plane are that displacements v and w and rotation θ_x are free, and that displacement u and rotations θ_y and θ_z are fixed to be zero. Figure 7.19 shows the coarse mesh with these boundary conditions imposed. Note the use of a double arrowhead to denote that the rotation about the axis direction has been fully restrained.

Now the computer model is complete and the solver can be run to produce the required results.

7.2.4 Finite Element Results

From our understanding of the requirements of this analysis, the results must be examined to determine the displacement of the vessel under the pressure loading and the maximum tensile stress within the material. So that any output can be easily visualized in the context of the vessel's geometry, the displacement results are presented for the deformed and undeformed shapes on one figure, and the stress results are given on the undeformed mesh. Analysts should not underestimate the time that it takes to generate the appropriate graphical pictures from the numerical data when using post-processors, as complex operations are often required to obtain, for example, a view of the geometry in the correct orientation or the correct levels for stress contours.

Figures 7.20 and 7.21 show the outline of the deformed shape together with the original coarse mesh, and fine mesh, respectively. Note that the deformed geometry is calculated from the magnitude and direction of the deflection at each node in the mesh, and that in these figures different magnifications have been used to show the deformed geometry. Without magnification the displacement from the original shape cannot be seen. From these figures the predicted deformation shows that the dome and the cap are forced outwards and that the shape expands along minor axis while contracting along the major axis. Hence the vessel becomess more circular in cross-section. This ties in with engineering intuition, which suggests that a thin-walled tube of elliptical cross-section would becomess more circular under a pressure loading, as energy arguments can be used to show that a circular section is the ideal shape for pressure loading. This also acts as a check on the modelling of the pressure vessel.

Figures 7.20 and 7.21 also show the maximum deflections in each of the Cartesian coordinate directions, together with the locations at which these deflections occur. By comparing these values and the overall deformation of each mesh, it is clear that the two meshes perform very similarly, with an error of

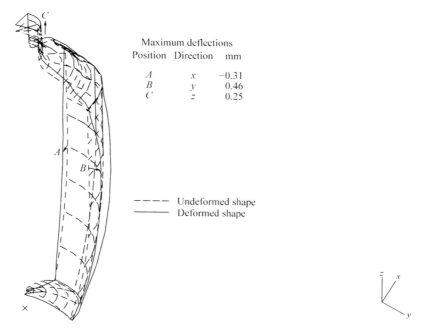

Maximum deflections

Position	Direction	mm
A	x	−0.31
B	y	0.46
C	z	0.25

- - - - Undeformed shape
———— Deformed shape

Figure 7.20 The displaced shape predicted with the coarse mesh.

approximately 3 per cent. Hence the h-refinement has shown that the original coarse mesh is suitable to this level of accuracy.

Note that the maximum deflection of the vessel body is 0.47 mm at point *B*, and is calculated with the fine mesh (Fig. 7.21). Owing to the restraints that have been applied, this means that the bottle increases in size by twice this value, i.e. 0.94 mm. As this is less than the specified limit for the increase of 1.0 mm, it can be concluded that the stiffness of the vessel is acceptable.

To consider the stresses within the model, the usual tool is a contour plot displaying a range of stress levels. In examining these plots the analyst can make some simple, yet valuable, checks. For example, the boundary conditions of the vessel require that the stress component normal to the surface must be equal to the pressure loading. Equally, at the planes of symmetry, the contour must be normal to the symmetry boundary. In general, there should be no abrupt changes in the direction of a contour and the contour should be continuous. However, where there is a change in material thickness, which in the vessel occurs at the boundaries of the 10 surfaces that were meshed, or where the geometry has a discontinuity, e.g. at the dome to side wall and body to shoulder interfaces, it is to be expected that the contour plots will be neither smooth nor continuous. Also, the contours should show the peak values of stress in the expected places.

Be aware that software packages do not always acknowledge the method used to calculate the nodal stress values which are then interpolated to create the

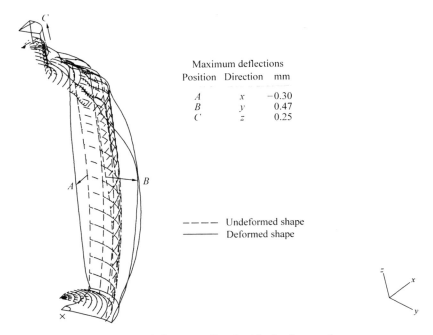

Maximum deflections

Position	Direction	mm
A	x	-0.30
B	y	0.47
C	z	0.25

– – – – Undeformed shape
———— Deformed shape

Figure 7.21 The displaced shape predicted with the fine mesh.

contour plots. For this reason these plots should never be used to find an accurate value of the peak stresses. If such values are required then the raw data, probably at Gauss points or possibly at the nodes, must be interrogated directly.

When generating a contour plot a commercial package may use default settings or an automatic data scaling routine to decide the intervals between each contour. Such an algorithm scans the nodal values to find the data range and then selects the contour intervals to cover the range of values for a given number of contour lines. Often the analyst has to override these defaults and set the range and intervals manually. From experience, plots with between 6 and 10 intervals give the best compromise between information provision and the clarity of display. If more than 10 intervals are used there is a danger that the plots become cluttered and difficult to follow. Equally, if fewer than 6 contours are used not enough information is presented and major features of the stress distribution may be missed.

Before presenting the stress contours for the vessel, the type of stress output available for a thin-shell element must be explained. Figure 7.22 shows that there are four direct stress values available. The element deforms due to both membrane (i.e. the middle surface stress σ_{ms}) and bending (i.e. the bending stress σ_B) actions. Note that for the isotropic shell element the bending behaviour is that of pure bending (see Sec. 2.3.1), where normals remain straight and normal. Combining the membrane and bending stresses gives two surface stresses, σ_{ts} at the top and σ_{bs} at the bottom.

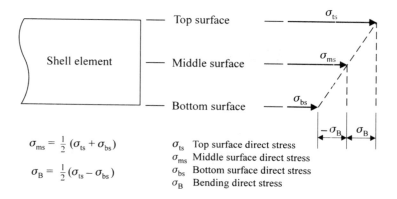

$$\sigma_{ms} = \frac{1}{2}(\sigma_{ts} + \sigma_{bs})$$

$$\sigma_B = \frac{1}{2}(\sigma_{ts} - \sigma_{bs})$$

σ_{ts} Top surface direct stress
σ_{ms} Middle surface direct stress
σ_{bs} Bottom surface direct stress
σ_B Bending direct stress

Figure 7.22 Bending stresses within shell elements.

As stresses are referred to the global Cartesian coordinate system, when selecting a principal stress (maximum, middle or minimum) the post-processor may calculate the three principal values from the displacement data. Values are determined to represent the top and bottom surfaces, as the post-processor assumes that the stress on the mid-plane is an average of the stress at the top and bottom surfaces.

Figures 7.23 to 7.31 are contour plots of the maximum principal stress, with the results for the coarse mesh being given in Figs 7.23–7.28 and those for the fine mesh in Figs. 7.29–7.30. Each plot gives the peak tensile and compressive stress, and the contour interval has been set manually to $5\,\text{N}\,\text{mm}^{-2}$. Note that the default settings in packages may lead to stress values for a contour being given to seven significant figures. Such use of output formatting could fool an inexperienced analyst into believing that this is also the accuracy for the stress output. The various procedures used by the post-processor generate contours from the raw data. This is known to introduce, in addition to those already inherent in the finite element analysis, further numerical errors. Hence, contour levels should not be given to more than three significant figures. In fact, in most cases, it is advisable to use only two significant figures.

From our understanding of the requirements of this analysis, if creep behaviour is to be neglected in design, then the direct stress must be less than $20\,\text{N}\,\text{mm}^{-2}$ in the body section of the vessel. Furthermore, it is expected that the peak values will be at the outer surfaces of the shell because of the combined membrane and bending actions, and so Figs 7.23 and 7.24 are plots of stresses σ_{ts} and σ_{bs} for all the elements in the model. From these figures it can be seen that the contours are not smooth or continuous in the regions of the vessel where the thickness changes or there are discontinuities in the shape of the vessel. Away from these regions, in the body of the vessel, it is found that the contours are fairly smooth with only slight rippling. It is only in such areas that the predicted stress

Contour	Direct stress σ_{ts} N mm^{-2}
1	−5
2	0
3	5
4	10
5	15
6	20
7	25
8	30

Peak values	
Compressive	39
Tensile	30

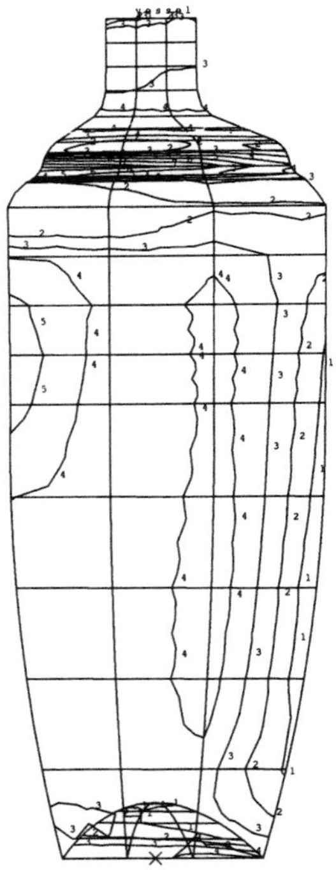

Figure 7.23 Maximum principal stress at the top surface (coarse mesh).

values can be used reliably, and inspection of the stress contours in the body region, away from discontinuities, shows that the surface stresses only just exceed the limiting stress of $20\,\text{N}\,\text{mm}^{-2}$.

To remove the influence on the generation of the contour plot of the dome, the shoulder and head of the vessel, the elements in these regions are not included in the post-processing for Figs 7.25 and 7.26. In Fig. 7.25 the stress at the top surface is shown, and so this figure can be compared to Fig. 7.23. It can be seen that the contours have changed only close to where elements have been removed. By limiting the stress contour plots to only those elements where the material thickness is constant, the effect of geometry changes is minimized. This same procedure of isolating specific parts of the structure when producing contour stress plots is also carried out if there are dissimilar material types or an abrupt change in the structural loading, for the same reasons.

When comparing Figs 7.25 and 7.26 to Figs 7.23 and 7.24, it can be seen that in the middle region of the body there is hardly any difference. Here, Saint-Venant's principle is at work and the 'die-away' length for this example is several times the wall thickness of 1.8 mm. For the reasons already discussed, the stress contours are not expected to be accurate close to the two interfaces and so the high stress values there should be considered as inaccurate.

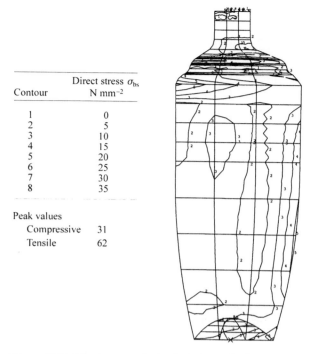

Contour	Direct stress σ_{bs} N mm^{-2}
1	0
2	5
3	10
4	15
5	20
6	25
7	30
8	35
Peak values	
Compressive	31
Tensile	62

Figure 7.24 Maximum principal stress at the bottom surface (coarse mesh).

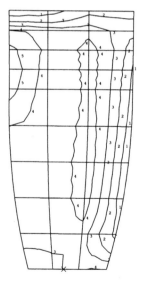

Contour	Direct stress σ_{ts} N mm^{-2}
1	−5
2	0
3	5
4	10
5	15
Peak value	
Tensile	19

Figure 7.25 Maximum principal stress at the top surface—body only (coarse mesh).

Contour	Direct stress σ_{bs} N mm^{-2}
1	5
2	10
3	15
4	20
5	25
6	30
Peak value	
Tensile	34

Figure 7.26 Maximum principal stress at the bottom surface—body only (coarse mesh).

New information on the stress distribution is provided in Figs 7.27 and 7.28. In Fig. 7.27 the middle surface stress σ_{ms}, known as the membrane stress, is fairly constant within the body having a value between 5 and 10 N mm^{-2}. In contrast, from Fig. 7.28, the bending stress σ_B continually varies from values below -5 N mm^{-2} to values greater than 25 N mm^{-2}. This shows that for the vessel the bending action is dominant over the membrane action.

Contour	Direct stress σ_{ms} N mm^{-2}
1	5
2	10
Peak value	
Tensile	13

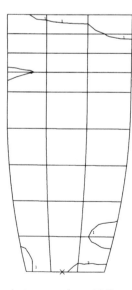

Figure 7.27 Maximum principal stress at the middle surface—body only (coarse mesh).

Contour	Direct stress σ_B N mm^{-2}
1	−5
2	0
3	5
4	10
5	15
6	20
7	25

Peak values	
Tensile	28
Compressive	8

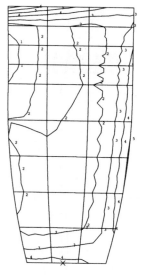

Figure 7.28 Maximum principal stress due to bending—body only (coarse mesh).

Stress results for the fine mesh are shown in Figs. 7.29 to 7.31. By comparing these figures with their coarse mesh equivalents it can be seen that, if anything, the contours are less smooth and have more discontinuities. It is, however, reassuring to find that the peak stress values are little different. This indicates that it is not necessary to run a third model with a mesh which has been generated by further h-refinement. Hence the results generated with both the coarse and the fine mesh models can be taken as being representative. This says nothing, however, about the worthiness of the results with regard to what actually happens.

7.2.5 Further Modelling Possibilities

The specification for the design of the vessel has been limited to the serviceability states of limiting maximum deformation and working tensile stresses due to the vessel's pressure charge. Analyses are also needed to model such things as impact behaviour. For example, a critical situation can be identified when a free-falling vessel impacts a hard surface and the point of contact is the interface between head and body. A linear elastic small displacement analysis can still be used if the impact load is modelled as an equivalent static load. From experience, such an impact load can be modelled as a static load which acts normal to the point of contact and the magnitude of which is twice the dead weight of the vessel. As a consequence of this the models defined here can be used providing that the boundary conditions are modified.

The causes of ultimate failure of the vessel may also need to be known. To do this a full nonlinear analysis must be carried out as the polymer material does not

Contour	Direct stress σ_{ts} N mm^{-2}
1	−5
2	0
3	5
4	10
5	15
6	20
7	25

Peak values	
Tensile	30
Compressive	38

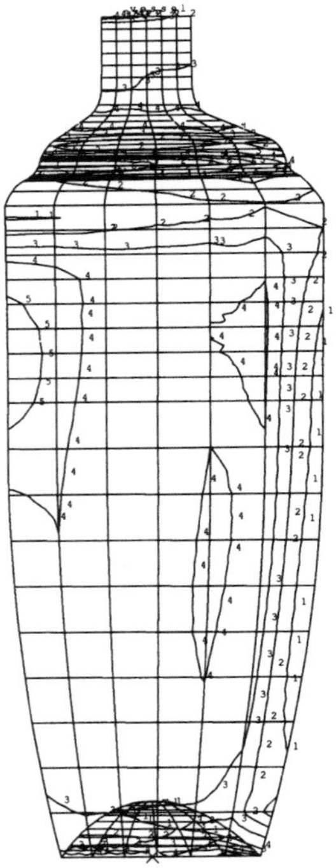

Figure 7.29 Maximum principal stress at the top surface (fine mesh).

Contour	Direct stress σ_{ts} N mm^{-2}
1	−5
2	0
3	5
4	10
5	15

Peak values	
Tensile	20
Compressive	10

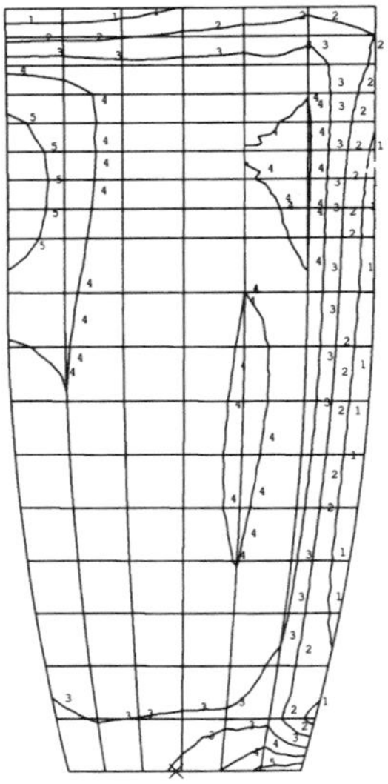

Figure 7.30 Maximum principal stress at the top surface—body only (fine mesh).

Contour	Direct stress σ_{bs} N mm^{-2}
1	5
2	10
3	15
4	20
5	25
6	30
7	35
Peak value	
Tensile	37

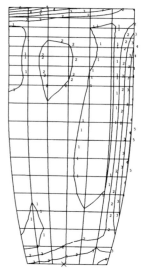

Figure 7.31 Maximum principal stress at the bottom surface—body only (fine mesh).

behave in a linear elastic way when failure occurs, as discussed in Sec. 2.2.6, as the peak deformations exceed the wall thickness and material properties are nonlinear. It may even be necessary to carry out a full nonlinear analysis for the impact condition if the deformations and strain rates are found to be large.

In all these nonlinear cases, global models may be used initially to identify the locations where failure is likely to occur. If such locations are near to discontinuities in the geometry or thickness, then a shell model will not provide accurate results that can be used to determine the loading causing failure. The analyst then has to decide if a model of local regions of the vessel is necessary. This may be achieved using the substructuring technique, explained in the frame example in Chapter 8, where the three-dimensional solid geometry of a local region is modelled using boundary conditions from the results of a global model. This is the only technique that can be used to determine the stress variations at the discontinuities in a shell model. However, it is not the panacea that the designer desires unless the exact material property variation and exact boundary conditions are known and can be modelled.

So it can be seen that the designer constantly wants the most accurate finite element results from the resources available, in terms of funding, the analysis team, software and hardware.

8

MORE ADVANCED PROBLEMS

Both of the examples presented in Chapter 7 might be considered simple problems as the material considered was isotropic and homogeneous, and a single type of mesh was used to analyse each structure. In this chapter the use of more advanced modelling techniques will be demonstrated, although still with linear static analysis. First, a beam made of a laminated composite material will be considered and then a steel frame, which will be analysed using the substructuring technique. In both examples parts of their structure are subjected to compressive stresses whose magnitudes have been checked to verify that no buckling instabilities will occur and so this aspect of structural design will not be considered further.

8.1 COMPOSITE MATERIAL OPEN-SECTIONED BEAM

The purpose of this example is to illustrate the use of commercial software in providing solutions for the class of engineering structures where the structure is made from laminations of materials that have anisotropic properties. These advanced composite materials have been developed over the last 30 years to have mechanical and other properties that make them attractive alternatives to the conventional structural materials such as metal alloys.

The word 'composite' is used, strictly, to describe a wide range of engineering materials whose essential characteristic is that there is more than one distinct material phase. Consequently, virtually all materials are composites. For example, in a metal alloy the dispersion of two of more phases is brought about by careful heat treatment. Here, however, the discussion of composites will be restricted to two-phase materials having discrete, long, cylindrical inclusions (or continuous fibres) in a continuum of a second phase or matrix. The fibres, $2\,\mu m$ or less in diameter, are usually of carbon, glass or aramid, although ceramic and metal fibres are being introduced, whereas the matrix can be of a polymer resin, metal or ceramic material.

Here the discussion will also be restricted to those materials of laminated construction where different layers are made up of different or similar composite material with dissimilar orientation of the fibres. Such multi-layered materials, often referred to as as laminated composites, consist typically of between 16 and 120 laminae. These materials were initially used in the structures of high-performance aircraft exploiting their high strength-to-weight ratio, but they are increasingly being used in structural applications across all engineering industries.

Before addressing the actual beam problem, the ways in which these laminate composites are built up must be considered, as well as the ways in which the finite element method is adapted to model their material properties. Only after this introductory material can the problem be described and a finite element model be built and analysed.

8.1.1 Properties of Laminate and Laminated Materials

Figure 8.1 illustrates a unidirectional fibre-reinforced lamina which is the basic building block of a multi-layered laminate (Halpin, 1992). The material of the lamina is assumed to be homogeneous and orthotropic, i.e. there are only three

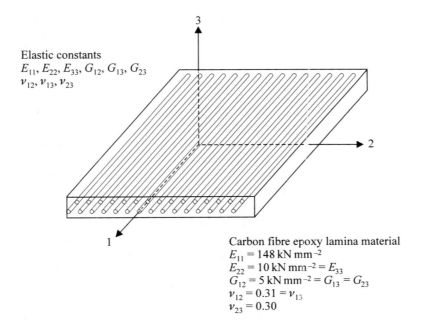

Elastic constants
$E_{11}, E_{22}, E_{33}, G_{12}, G_{13}, G_{23}$
$\nu_{12}, \nu_{13}, \nu_{23}$

Carbon fibre epoxy lamina material
$E_{11} = 148 \, \text{kN mm}^{-2}$
$E_{22} = 10 \, \text{kN mm}^{-2} = E_{33}$
$G_{12} = 5 \, \text{kN mm}^{-2} = G_{13} = G_{23}$
$\nu_{12} = 0.31 = \nu_{13}$
$\nu_{23} = 0.30$

Figure 8.1 Schematic representation of a unidirectional composite, showing principal axes.

mutually perpendicular planes of symmetry of material properties at any one point. Also, it is assumed that there is a perfect bond between the fibres and matrix, that the composite has linear elastic properties, and that the elastic behaviour of the material in tension and compression is the same. It is often a good assumption to assume that stress–strain relationships are linear elastic up to ultimate failure, as shown in Fig. 2.6(d).

By convention, the principal axes of a lamina are labelled 1, 2 and 3 with the 1-axis parallel to the fibres. Since there is isotropic symmetry in the 23-plane which is normal to the fibres, a unidirectional reinforced lamina is a transversely isotropic material. Hence six independent elastic constants (i.e. E_{11}, E_{22}, G_{12}, G_{23}, ν_{12}, ν_{23}) must be used to generate the terms for the full three-dimensional form of Hooke's law, Eq. (2.15). These constants are given in Fig. 8.1 for the carbon fibre epoxy material used here. Note that the Es are the principal Young's moduli and the Gs are the principal shear moduli. Also note that for the lamina considered here the subscripts on the stresses and strains in (2.15) need to be changed from x, y and z to 1, 2 and 3, respectively.

Note that there are problems when obtaining material property data for a unidirectional fibre-reinforced lamina material. Owing to anisotropy many standard coupon tests (see Sec. 2.2.6), are needed to determine the six elastic constants and six ultimate strengths. As there are different modes of failure in compression and in tension, these strengths, along and normal to the fibres, are very different.

Table 2.1 gives a range of mechanical properties for unidirectional materials with a polymer matrix. This data can be obtained only after the material has been manufactured and it depends on the volume fraction of fibre, fibre placement and the method of manufacture. Equally, many combinations of fibres and matrix are available and new combinations are appearing all the time, so it is difficult for an analyst to have data for all the materials available. Hence, analysts must check any lamina data and use only values that will ensure a safe design.

Figure 8.2 shows a general laminate which consists of four unidirectional laminae of constant thickness. Here, the fibre directions are not aligned to the principal loading direction, the local laminate x-direction (i.e. the 0 degree direction). To identify the lamination configuration, a notation derived from the fibre directions, i.e. $(-45/30/30/-45)$, is used. In practice, there are several types of lamination configuration and a number of these are used in the example.

When a lamina is orientated at a general angle to the local laminate x-direction of the laminate, the generalized Hooke's laws (2.15), for the layer are obtained using standard axis transformations. Note that the stresses and strains are now in terms of the Cartesian axis system and the terms in matrix [**D**] are functions the lamina stiffness terms. Once the stress–strain relationship for the lamina is known together with its position in a laminate, it is possible to form the relationships between the stress resultants due to axial forces and bending moments and the membrane strains and plate curvatures respectively (Halpin, 1992).

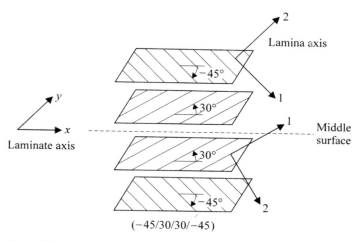

Figure 8.2 An example of notation for a laminate.

8.1.2 Laminated Composites and the Finite Element Method

Here, only a brief introduction to the ways in which the finite element method is used to model anisotropic laminated material properties will be given. Laminated composites are usually represented by a series of equivalent laminated homogeneous plates or shells. Here, the standard form of the element (see Chapter 3) is used, but with the appropriate laminate properties constructed from a compact description of the laminate stacking sequence and lamina material properties, as discussed at the end of Sec. 8.1.1, where relationships for the axial, coupling and bending stiffnesses are produced.

To develop the element properties, the pre-processor is used to define the elastic constants of a lamina and the configuration of the lamination. Then, within the solver, the stiffness matrices for the laminate can be calculated, and assigned to the elements. Hence, when dealing with a laminated composite material as opposed to an isotropic material, it is only the material properties, i.e. the terms in matrix [**D**], that change when element stiffness matrices are formed using (3.9). Note that for all element types, laminate properties may introduce coupling effects between membrane and bending actions that radically change the deformation from what might be expected if the material were isotropic. These coupling effects must be included in the analysis.

Laminated thin-shell elements, both linear and quadratic, are often formulated using approximate solutions to the Mindlin shell theory. This accounts for shear deformation by forcing the normals to a shell to remain straight but then allowing them to rotate such that they are not normal any more. Note that the transverse shear actually causes a normal to become curved. When deformation takes place, if the normals are assumed to remain straight and normal to the surface, which is Kirchhoff's assumption, the laminated shell element is formulated using classical lamination theory. However, as the inclusion of shear deformation is usually necessary when analysing laminate structures, many solvers do not use classical lamination theory elements at all.

Once nodal displacements have been calculated, the stresses and strains in individual layers have to be recovered and resolved into the appropriate material axes. This data is used to determine a number of layer-by-layer failure criteria, and these are strongly influenced by the method of stress recovery. For laminates this is often the weak link in the analysis. Interpretation of any failure prediction is further hampered by the lack of any general consensus as to the best criterion to use. It is for this reason that failure predictions will not be presented here.

Some solvers allow laminates, and other layered composites, to be constructed of so-called stackable plate and solid elements. This allows a partial three-dimensional analysis to be carried out at a fraction of the cost of a full layer-by-layer analysis with solid elements.

8.1.3   Problem Specification

A symmetrical open-section I-beam of length 1600 mm is to be tested in three-point bending; the dimensions of the I-shaped cross-section are given in Fig. 8.3(a). The web and flanges are formed from laminate panels 2 mm thick and consist of 16 laminae of a unidirectional carbon fibre epoxy high-modulus material. Three different lamination configurations for web and flange panels are used and these are detailed later.

The loading arrangement is shown in Fig. 8.3(b), where simple supports are placed 1600 mm apart and a central vertical load of 1 kN is applied to the top

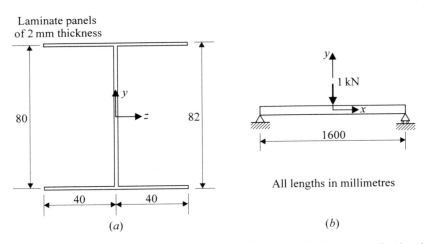

Figure 8.3 The composite material beam problem: (*a*) the I-section, (*b*) the three-point loading arrangement.

flange. This load has been chosen to ensure that any displacements of the beam are small. As well as this configuration, spans of 400, 800 and 1200 mm are also analysed.

The above situations are analysed for the following reasons:

- To illustrate how the deformation of the cross-section is affected by the coupling that can exist between bending and twisting, and membrane and bending actions, when different types of lamination configurations are used.
- To show typical strain and stress output from a finite element analysis for a laminated composite panel.
- To determine the section flexural modulus, and the section shear modulus of the I-section. Note that a given section modulus is not necessarily the same as the corresponding material modulus, as the former will be determined from a theoretical expression for deflection, while the latter is determined using standard material coupon test methods. Both section properties will be found using an extension of beam theory developed in the next section, together with a graphical procedure.

8.1.4   An Extension of Simple Beam Theory

In Sec. 8.1.2 it was stated that laminated structures are often modelled with laminated shell elements formulated using Mindlin shell theory. This is because of the important role that material properties play in determining the deformation of a structure, such that a theoretical model which is acceptable for an isotropic material may not be acceptable for an anisotropic material. Analysts must, therefore, plan a finite element analysis with due regard to the likely deformation behaviour and, when developing the model, choose the best element available.

To illustrate why, for example, shear deformation is not to be ignored when using laminated composites, consider again the simple bending problem of a prismatic beam, introduced at the end of Chapter 2. There, (2.25) gives the general relationship used to analyse bending. It is derived by assuming that the properties of the beam can be lumped along the centroidal axis and that when bending occurs Kirchhoff's assumption holds. For a small displacement analysis, the expression

$$R = \pm \frac{1}{\dfrac{\mathrm{d}^2 v}{\mathrm{d} x^2}} \tag{8.1}$$

where R is the radius of curvature, v is the deflection of the beam in the y-direction at the centroidal axis and x is the local coordinate along the beam's centroidal axis, can be used to form the well-known moment–curvature relationship

$$\frac{M(x)}{EI_\mathrm{a}} = \pm \frac{\mathrm{d}^2 v(x)}{\mathrm{d} x^2} \tag{8.2}$$

Expressions for the deflection at any position along the beam can be obtained using (8.2) if the variation in bending moment along the beam is known together with the boundary conditions for deflection and its slope.

For the three-point bending problem shown in Fig. 8.3, the maximum deflection due to bending is at the mid-span and is given by the well-known expression

$$v_b = \frac{PL^3}{48E_b I_a} \tag{8.3}$$

where P is the applied load, L is the span and E_b is the beam section flexural modulus. Here the negative sign has been ignored for convenience.

Except for the situation where the beam is loaded by a pure moment, shear forces are always present and are transmitted through the cross-section of the beam. These shear forces cause shear stresses that give rise to an additional deflection, known as shear deformation. It can be shown using linear elastic assumptions that this second deflection component is also maximum at mid-span and is given by

$$v_s = \frac{PL}{4K_s G_b A} \tag{8.4}$$

where K_s is the shear correction factor that accounts for the shear stress distribution not being constant over the cross-sectional area, G_b is the beam section shear modulus and A is the area of the cross-section. Values for K_s are available in the literature for beams with isotropic or orthotropic properties and for a number of cross-section shapes. However, to simplify (8.4) for the purpose of engineering design with I-sectioned beams, it is acceptable to assume that $K_s \equiv A_w/A$, where A_w is the area of the web, implying that the shear force is transmitted by the web alone and that the shear stress distribution is constant.

Combining the deflection components from bending and shear, the maximum deflection is given by

$$v_{max} = v_b + v_s = \frac{PL^3}{48E_b I_a} + \frac{PL}{4G_b A_w} \tag{8.5}$$

It is usually acceptable, however, when considering isotropic beams to ignore the shear deflection component. To see why, consider the relative magnitudes of the two deflection components. Looking at the terms in (8.5), it can be seen that

$$\frac{v_s}{v_b} = \frac{12E_b I_a}{G_b A_w L^2} \tag{8.6}$$

If the material is isotropic, then the ratio E_b/G_b (or E/G) is 2.6. For this case the shear deformation is small compared to the bending deformation, provided that the beam has a span-to-depth ratio of greater than 5. For typical steel beams this is the case. If not then the beam is termed *stocky* and shear deformation must be accounted for. The situation is, however, very different for an identical beam made of laminated composite material, as E_b/G_b is much greater than 2.6, possibly 30 or more. Hence, from equation (8.6), v_s cannot be ignored. As a rule of thumb the shear deformation contribution to v_{max} must be included for a laminated composite beam when the span-to-depth ratio is less than 25.

In Fig. 8.3, the span-to-depth ratio for the beam is 19.5, but this decreases to 4.9 when the span is 400 mm. For all cases, therefore, the shear deformation behaviour is expected to be significant and so this beam problem cannot be modelled using standard one-dimensional beam elements, as would be the case if material were isotropic.

8.1.5 Defining the Finite Element Model

Before beginning to build a model of the beam, the problem must be considered a little more. To take advantage of any symmetry in the problem a suitable choice of coordinate system and origin is required. Figure 8.3 shows the global system, where the x-direction is along the span of the beam, the y-direction is in the vertical direction and the z-direction is horizontal across the flanges. For this system, the origin is placed at the mid-span in the x-direction and at the centroid of the cross-section (i.e. at mid-depth of web).

There are also certain expectations of the deformation of the beam when subjected to three-point bending. Applying symmetry arguments, the cross-section at mid-span will be restrained such that it moves only in the z–y plane, i.e. it is a plane of mirror symmetry. If the material is isotropic and any displacements are small, then from Kirchhoff's assumption the normals will remain straight and normal, such that any cross-section remains planar. In this situation (2.25) holds and so, given the symmetry in the x-direction, only one-quarter of the beam need be modelled. However, sections do not necessarily remain planar because of shear deformation and coupling effects, as discussed in the last section, and so the model will be for the half-beam about the mid-span. Finally, if the mid-span plane is retrained to move in the y–z plane, then one of the simple supports must be free to move in the x-direction to prevent membrane action being induced in the flanges.

Geometry and mesh description   The geometry of the physical beam is shown in Fig. 8.3. As the panels have a thickness of only 2 mm relative to a minimum width of 40 mm, they can be considered to be thin plates. This leads us to use elements known as thin laminated shell elements to model the beam. These elements are placed such that they are at the mid-plane of the panels, and so care is needed to ensure their correct location. Each of the three rectangular panels of the beam must be modelled as shown in Fig. 8.3, but the depth of the model will be only 80 mm to account for the 'zero' thickness of the elements. Only when the laminate properties are defined for the elements will the model simulate the correct depth of the beam. Note also that quadratic elements are used because they provide the best solution of all the element types available.

Meshing the beam is relatively simple in this case as there are three flat areas to mesh, each 800 mm long and 80 mm wide. Lines in three-dimensional space can be used to define simple flat surfaces for each side of the upper flange and the upper half of the web, and these surfaces are meshed with a run of 20 quadrilateral elements which are 40 mm square. There will be no distortion from the parent element shape. This mesh can then be mirrored in the x–z plane to produce the final mesh of the half-beam. This mesh is shown in Fig. 8.4 and consists of 120 elements, 413 nodes and 2478 nodal degrees of freedom.

Materials   As the element type has been specified as a shell, the laminate material properties must be assigned to each element. For each element there is a local coordinate system which, as the elements are essentially planar, is just x and y. Here the local x-direction for a laminate element coincides with the global x-direction of the beam, as shown in Fig. 8.4. These laminate properties consist of axial, coupling and bending stiffnesses which are generated by the pre-processor using a laminate modelling facility. The user has, typically, two options to produce the lamina elastic constants, either entering them directly or entering the fibre and matrix constituent properties and allowing the software to calculate the constants using micromechanical formulae. Then the user supplies the orientation and thickness of each lamina in the stacking sequence, starting at the bottom layer and working to the top layer.

In this case, the former method was used to enter the unidirectional carbon fibre epoxy lamina data given in Fig. 8.1. Then 3 different examples of laminates, each 2 mm thick with 16 layers of this lamina, were constructed. Their laminate

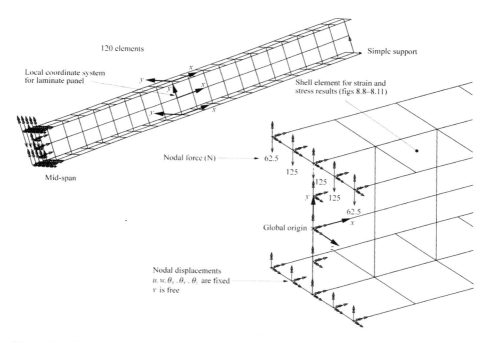

Figure 8.4 Mesh construction and boundary conditions.

Table 8.1   Examples of lamintes used for the beam problem

Label	Lamination notation/type	Description
Laminate 1	(0/90/0/90/0/90/0/90/90/0/90/0/90/0/90/0)	Symmetrical cross-ply
Laminate 2	(0/ − 45/90/45/0/ − 45/90/45/45/90/ − 45/0/45/90/ − 45/0)	Symmetrical quas-isotropic
Laminate 3	(0/ − 45/90/45/0/ − 45/90/45/ − 45/0/45/90/ − 45/0/45/90)	Antisymmetrical quasi-isotropic

configurations are defined in Table 8.1. Clearly, a large number of laminate configurations could have been used. These three configurations have been chosen because each represents a specific lamination type with a certain deformation behaviour. In reality, it is not easy to produce a beam with any general lamination configuration. One practical option is to manufacture a laminated I-section by joining two 8-layer channel sections back-to-back and then adding one 8-layer plate to each flange to achieve the appropriate thicknesses. Such an option may be used to produce the laminate configurations in the table.

Considering each of the laminate configurations in turn, the following may be seen:

- Laminate 1 has an equal number of $0°$ and $90°$ orientation laminae, stacked symmetrically about the mid-plane. Hence, the laminate stiffnesses for a thin-shell element are those of an equivalent orthotropic material, i.e. there is no coupling between bending and twisting and membrane action and bending. It can therefore be expected that a beam made of this material will behave in a way similar to an isotropic beam.
- Laminate 2 is a quasi-isotropic laminate, as it has an equal number of layers with orientations of $0°$, $90°$, $45°$, and $−45°$. It also has a stacking sequence that is symmetrical. Such a laminate, when produced with carbon fibre reinforcement, is known as *black aluminium,* because all its stiffnesses do not change significantly with angle. The presence of the $45°$ laminae introduce bending coupling stiffness terms that cause the material to bend as it twists and twist as it bends. However, when such a panel is subjected to membrane deformation there is no coupling with bending deformation. This feature of the deformation is due to the symmetrical stacking sequence.
- Laminate 3 has the same layer orientations as laminate 2 and is of the quasi-isotropic type. However, the stacking sequence about the mid-plane is what is termed 'antisymmetrical'. Now the laminate has coupling between membrane and bending deformations, but the terms which denote bending–twisting coupling are zero.

Loads and restraints   Remembering that the load applied at the mid-span is 1 kN (see Fig. 8.3), and that symmetry has been applied such that only half of the beam is modelled, only 500 N need be applied at the mid-span for the model. This load can be assumed to be uniformly distributed across the upper flange. To calculate the load that is applied at each node, consider the length over which the load acts. Figure 8.4 shows that the inner nodes along the line at the mid-span of the upper flange act over twice the length when compared to the edge nodes, and so twice the load is applied to the inner nodes. Hence, the two nodes at the free edges have point loads of -62.5 N in the y-direction applied to them and the three internal nodes have points loads of -125 N in the same direction applied to them. Here, the minus sign indicates that the load is acting in the negative y-direction. Note also that the arrows which denote the loads shown in Fig. 8.4 are all of one length and so the graphical display cannot be used to check the accuracy of the model in this case.

Applying the correct restraints to the nodes that are on the mid-span plane of the beam must ensure that the nodes stay in the y–z plane and move only in the y-direction. This involves allowing only the v-displacement to be free, while setting the u- and w-displacements and the θ_x, θ_y and θ_z rotations to be zero.

To model a simple support, the nodal displacement boundary conditions can be one of two types. If the deformation is such that sections remain planar, then all the nodes across the width of the bottom flange at the support location have their nodal displacement u and rotation θ_z set to be free. A value for u set to zero represents the situation where axial shortening of the beam is not allowed, and this has not been included when deriving (8.5). If sections do not remain planar, i.e. there is twisting of the section which causes warping of the section, it can be observed that the rotational nodal displacements at the support may be nonzero. To model both cases, the node at the bottom of the web has its nodal displacements v and w fixed to be zero, while all other nodes across the width of the bottom flange have their degrees of freedom left free.

In Fig. 8.4, double-arrowheaded vectors are used to show those nodal displacements that are fixed. This can be used to check whether the conditions imposed on the model are those required.

Analyses   For each of the three laminates in Table 8.1 models have been built, as described, to represent spans of 1600 mm. Also spans of 400 mm, 800 mm and 1200 mm have been modelled by moving the position of the simple support. Linear static analyses have been carried out for each case.

8.1.6   Results of the Analysis

Once an analysis is run the raw data is checked to make sure that the stresses and displacements are acceptable. This is done visually using a graphics terminal and, for the cases run here, both the stress and displacement values are found to be small. Hence, the underlying assumptions of small displacements and linear elasticity are seen to be valid. Now, the results can be analysed in more detail.

Displacement contour plots   To illustrate how the deformation of the beam alters with different laminate configurations, Figs 8.5 to 8.7 present displacement contours for the three laminates on the model of half-beam span. The contours are of constant v-displacement, i.e. in the global y-direction, and are drawn on the deformed geometry so that the global deformation is visible. For reasons of clarity the undeformed geometry and the mesh are not shown here. However,

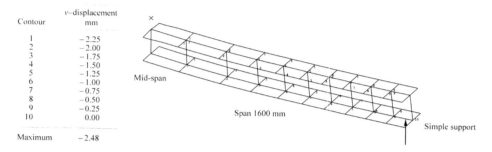

Figure 8.5 Contours of v-displacement on the deformed beam with laminate 1 (cross-ply).

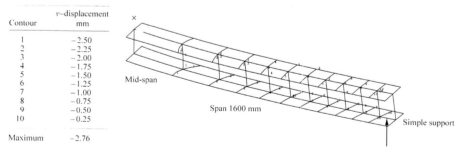

Figure 8.6 Contours of v-displacement on the deformed beam with laminate 2 (symmetrical quasi-isotropic)

Contour	v-displacement mm
1	-2.50
2	-2.25
3	-2.00
4	-1.75
5	-1.50
6	-1.25
7	-1.00
8	-0.75
9	-0.50
10	-0.25
Maximum	-2.76

Figure 8.7 Contours of v-displacement on the deformed beam with laminate 3 (antisymmetrical quasi-isotropic).

to show that it is correct to allow the u-displacement to be free at the simple support, the original location of the support is shown. In all the figures, the contours have a uniform interval chosen by the analyst.

From the figures it can be seen that the displacement component v is negative as the beam deflects in the negative y-direction. In design the minus sign is ignored as it is the magnitude that is required, the direction being unimportant for safe design. Next, note that the contours are not uniformly spaced along the beam; they become closer together when moving from the mid-span to the end of the beam. This is indicative of a parabolic variation in v along the beam, which is expected from the moment–curvature expression.

If plane sections of the beam remain plane under loading, then this is shown by the displacement contours being undistorted and at right angles to the centroidal axis of the beam. For the beam made from laminate 1, Fig. 8.5, this is the case as expected. Laminate 2, however, should lead to some twisting, but this cannot be seen in Fig. 8.6. If the raw nodal displacements are interrogated, then some slight twisting of the flanges is found, but that the magnitude of the twisting is too small to be detected from the contour plot. This shows that relying exclusively on visual interpretation of the results can lead to incorrect answers. Figure 8.7 for the beam made with laminate 3 is very different from the other two, as significant distortion of the cross-sections of the beam is visible. Comparing Fig. 8.7 for an antisymmetric laminate with Fig. 8.6 for a symmetric laminate, the membrane-bending coupling of the antisymmetric laminate can be seen to have a large influence on the deformation behaviour. For this reason, in real structures, antisymmetric laminates are not normally used, unless their membrane-bending coupling is desirable.

Also shown in each figure is the maximum vertical deflection. The beams made from laminates 2 and 3 have a similar value, showing that the strong coupling effects do not affect the maximum bending deflection. This has important consequences as (8.5), which has been derived by lumping properties of the beam along the centroid axis, as introduced in Sec. 2.3.1, can be used to determine v_{max} for any laminated composite beam subjected to three-point bending. This does not mean, however, that any coupling effects present in the beam can be neglected in the overall design process, because they will undoubtably have an influence when designing such things as joints and supports.

Laminate stress and strain output   Having looked at the global deformation of the structure, the stress and strain within the laminate material itself must now be analysed. This is done within a specialist post-processing facility designed to handle multi-layered materials. This facility combines the nodal displacements for the thin-shell elements that represent the laminate, generated by the solver, with the laminate stiffnesses, laminate configuration and lamina elastic constants, to calculate layer stresses, with reference to the local Cartesian coordinate system of the laminate.

Analysts must expect that this process cannot generate accurate results for the individual layer in-plane stresses, σ_x, σ_y and τ_{xy}, and the interlaminar stresses, τ_{xz}

and τ_{yz}. These last two shear stresses are often estimated only approximately, a common method being to use simplified layer-by-layer equilibrium equations. Note also that, normally, the laminate post-processor does not generate output for σ_z the through-thickness stress, known as the *peel stress*. As laminates are susceptible to delamination damage, caused by such stresses, this lack of information is very serious. These comments on accuracy also apply to the recovery of layer strains.

Examples of the graphical output that can be obtained for a given thin-shell laminate element are shown in Figs 8.8–8.11 for a beam of span 1600 mm made for laminate 1. The stress and strain recovery has been carried out for the element marked in Fig. 8.4. These values are at the centroid of the element.

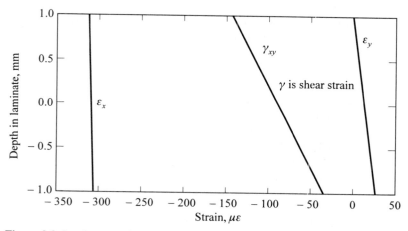

Figure 8.8 In-plane strain profile for the compression flange with laminate 1.

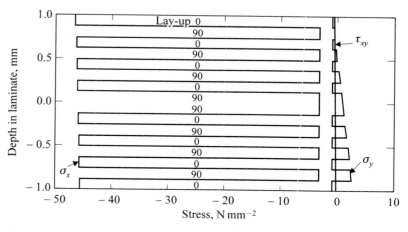

Figure 8.9 In-plane stress profiles for the compression flange with laminate 1.

In Fig. 8.8, the in-plane strain profiles are continuous and vary linearly through the material thickness. This shows that normals have remained straight, a feature of Mindlin shell theory (Sec. 3.8.2), and that the compatibility requirement is satisfied. Using (2.25), with the flexural modulus E_b for laminate 1 in Table 8.2, it can readily be shown that the top surface strain (ε_x) is $-300\,\mu\varepsilon$ and that the bottom surface strain is $-315\,\mu\varepsilon$, a difference of only $15\,\mu\varepsilon$. The finite element results are in good agreement with this strain due to bending about the horizontal $z-z$ axis. Such an observation indicates that the flanges resist the bending by membrane action, this being the general situation when the walls of an I-section beam are thin. The nonzero values for ε_y, the strain perpendicular to the longitudinal axis of the beam, show that there is the expected deformation in the plane perpendicular to the beam's length, as introduced in Sec. 2.3.1.

In-plane stresses are plotted in Fig. 8.9. Although the stress profiles vary linearly through the thickness of a lamina the values are different from layer to layer because of the different material properties in each layer. The stress σ_x is highest in the stiffer $0°$ layers of the cross-ply laminate and, as expected because of the inherent errors in the finite element model, the results for σ_y and τ_{xy} are not exactly zero, but oscillate around this value. In each layer, the in-plane stress distribution in each layer satisfies the equilibrium requirement.

Transverse shear strain and shear stress profiles through the top flange laminate are presented in Figs 8.10 and 8.11. Through thickness shear values are very small because the majority of the shear force is transmitted by the web. Both strain and stress profiles are close to the parabolic shape that is predicted by conventional elastic theory. The deviation of the results from the expected shape is a consequence of using simplified equilibrium considerations to calculate the shear values.

Having obtained the above profiles in Figs 8.8–8.11 it is usually possible to plot them on a layer-by-layer basis having transformed the laminate local axis strains and stresses to be in the principal axis system (see Fig. 8.2). These principal values are then used for the prediction of failure. Von Mises' equivalent stress is no longer relevant here, since failure of a lamina, and hence the laminate, is directionally dependent. The anisotropy of the material also tends to mean that

Table 8.2 Section moduli of beam compared to in-plane and flexural moduli of laminate

	Section		Laminate in-plane		Laminate flexural
	\multicolumn{5}{c}{Modulus values, $kN\,mm^{-2}$ (GPa)}
Laminate	E_b	G_b	E_x	G_{xy}	$E_{x,zz}$
1	79.8	4.46	77.9	5.35	92.5
2	57.6	20.8	63.0	21.6	80.7
3	57.1	18.2	63.0	21.6	66.0

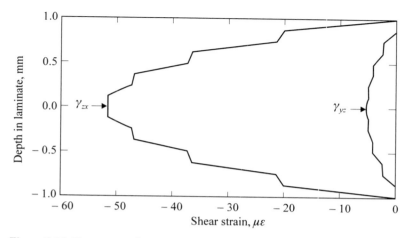

Figure 8.10 Transverse shear strain profiles for the compression flange with laminate 1.

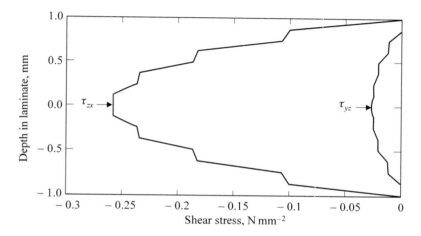

Figure 8.11 Transverse shear stress profiles for the compression flange with laminate 1.

the individual stress components are more likely to cause failure and this is one reason why post-processors have the options discussed here. However, note that there can be difficulties associated with using finite element analyses to predict such failures.

Often the actual stress causing failure is at a free edge of the material and the calculations are carried out at the centroid of the element. Also, it is not necessarily the largest stress that causes failure. It has been established that the error in any stress component is 5 per cent of the maximum stress component. This indicates that if a stress component causing failure is an order of magnitude less than the maximum component, then the error in this stress can be 50 per cent, and as such is not useful in predicting failure.

Determination of section moduli   When developing (8.5) for the mid-span deflection, it was stated that E_b and G_b were those of the section. To determine their values for a laminated composite I-beam there are a number of alternative procedures; one of these uses solutions from finite element analyses and will be demonstrated here.

Rearranging (8.5) gives

$$\frac{v_{max}}{PL} = \frac{1}{48E_bI_a}L^2 + \frac{1}{4G_bA_w} \tag{8.7}$$

which is the equation of a straight line if v_{max}/PL is plotted against L^2, with gradient $1/48E_bI_a$ and intercept $1/4G_bA_w$. For the beam section being considered here, the second moment of area I_a is $5.91 \times 10^5\,\mathrm{mm^4}$ and the web area A_w is $160\,\mathrm{mm^2}$. Hence by plotting v_{max}/PL against L^2 for various span lengths, the gradient and intercept can be found, giving the section moduli.

While the location of v_{max} may be at any point within the mid-span section, to be consistent with (8.5) the value of deflection at the centroid of the section has been used. Figure 8.12 shows the results from the beam made of laminate 1. Note that the four data points give an approximation to a straight line, and so the gradient and intercept are determined using the least-squares method to fit the best straight line through the data.

Results for the section moduli produced using this graphical method are given in Table 8.2 together with the equivalent axial moduli for the laminate panels. As the I-section is thin-walled, when the section is flexed the flanges are subject to mainly membrane action as the strain variation through their thickness because of bending is small, as shown in Fig. 8.8. It is not surprising, therefore, that E_x of the laminate is, for the purpose of design, a conservative approximation to E_b. Here, the difference is due to the actual location of the layers with respect to the axis of bending. Similarly, G_{xy} is a conservative approximation to G_b. Such a favourable

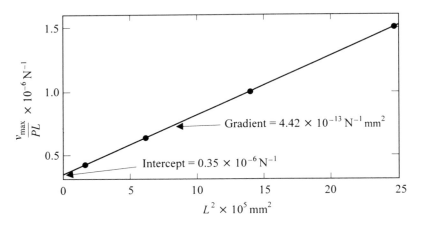

Figure 8.12 Determination of the section modulus with laminate 1.

comparison is not found if the section has thicker walls, and so it is always sensible to use the graphical procedure demonstrated here, with data either from a series of computations or from laboratory testing, to determine the section moduli.

The laminate itself also has its flexural moduli. One of these, $E_{x.zz}$ is the modulus in the x-direction when a single laminate is bent about the z–z axis. This flexural modulus is much higher than the equivalent flexural modulus for the section, and this comparison illustrates, yet again, the importance of material properties in the analysis of structures. Note that all three E values are the same for an isotropic beam—the Young's modulus of the material.

This I-beam example has introduced a number of features that are relevant when analysing structures made from laminated composite materials. There are many more features that can be studied, and those requiring additional information on the analysis of composite material structures are recommended to study the comprehensive work by Taig (1992). This NAFEMS publication is a very practical document, exploring all aspects of those commercial packages available at the time of publication, together with a discussion of their weaknesses and degree of applicability of their features.

8.2 DESIGN EXAMPLE WITH SUBMODELS

8.2.1 Overview of the Problem

In all analyses, the aim is to have as few elements and spatial dimensions as possible, so that both the overall cost of the analysis and the numerical manipulation errors are kept to a minimum. One way of doing this for complex problems is to carry out a variety of analyses using, where necessary, one-, two- and three-dimensional elements to determine the results required. To illustrate this, consider the design of a frame made up of thin-walled steel tube members, the ends of which are welded together to form the joints. This frame supports a vibrating machine, but as an approximation only time-independent loads are assumed to be transmitted through the joints into the frame.

When analysing such a frame, a first step is to model the skeletal structure as either a series of one-dimensional bars with pin joints or as a set of beam elements with fully rigid joints. The mesh is generated by hand and the range of member stresses and/or forces and of joint and member deflections in the frame are obtained from a series of analyses carried out with static loads applied. Joint behaviour, however, for this practical situation is not predicted in any detail. For example, in this case, the analyst may need to know the local stresses in the vicinity of the connections so that the welds can be designed to avoid a fatigue failure.

It can be seen that to model this situation the use of one-dimensional elements alone is not appropriate. Such a one-dimensional analysis may, however, still be useful if the information it generates can be used for further local analyses of the joints. This is an example of what is known as substructure (or subregion) modelling, as described by Knight (1993). Here, results from the frame analysis can be used as boundary conditions for the subsequent analysis of a joint, where both the joint and a length of each the members attached to it can be modelled with two-dimensional shell elements. This allows both in-plane and bending behaviour of the joint system to be modelled. If this second-level model does not provide stresses suitable for design purposes then a third-level model of solid three-dimensional elements, with or without shell elements, may be necessary.

Note that only through experience will an analyst know how to formulate the subregion models such that an appropriate number and types of element are used to determine the required deformations and stresses with the degree of accuracy necessary to meet the design requirements.

8.2.2 Problem Definition

This example is used solely to illustrate a number of features of the finite element method and so it does not correspond exactly to a real design situation. Hence, the type of frame and loading arrangement is not typical of those used in practice. However, property data for commercially available steel members has been used. Interested readers are referred to any of the available design guides for the construction of frames with hollow steel sections, such as that by Packer, Wardenier, Kurobane *et al.* (1992).

For this example, a space frame of nominal dimensions 3000 × 1500 × 1500 mm is constructed from steel sections. Members are chosen from a range of welded tubes of rectangular and square section. As is typical of these steel frames, the method of connection at the joints is welding and, to achieve full strength, full-penetration butt welding is used.

This frame is to support a static or dead load, a machine that weighs 72 kN (equivalent to about 7 tonnes), above ground level and so the base of the frame is attached to the ground. At this stage in design the dead load of the frame itself is unknown, but it is not expected to be more than 5 per cent of the machine's dead load. A number of components on the machine move such that the frame is subjected to dynamic loading in addition to the dead load. It is therefore one of the design requirements that the welded connections have an adequate fatigue life. To model the structural behaviour it can be assumed that loading is of constant amplitude and of low frequency such that the dynamic loading can be modelled in the finite element analyses by a quasi-static loading, thus enabling results to be obtained from a purely static analysis.

It is known that the motion of the components provides a dynamic magnification to the dead load of 1.25, making the maximum total vertical load 90 kN and the minimum load 54 kN. In addition, there is a horizontal dynamic load component equal to 0.2 of the dead load. Both dynamic induced loads act in the plane parallel to the longer side of the frame which is illustrated in Fig. 8.13. The dynamic loading perpendicular to this plane is found to be sufficiently small such that it can be neglected.

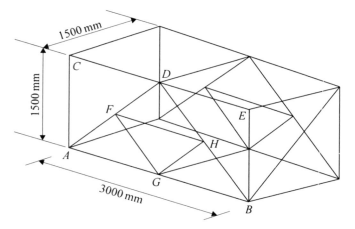

Figure 8.13 The nominal geometry of the frame.

Preliminary design studies show that the frame has the geometry shown in Fig. 8.13, where the joints on one side of the frame have been labelled to aid the following discussion. To simplify the design process advantage is taken of symmetry and so only the plane frame that is labelled and which is subjected to only half of the loading need be considered. In the preliminary design studies account has been made of the way in which the machine is mounted to the frame at points along the horizontal member *FH*. From this, the maximum quasi-static load distribution is known and is shown in Fig. 8.14. The vertical component of load is represented by a linearly varying distributed load of value $12\,\text{N}\,\text{mm}^{-1}$ at joint *F* and $48\,\text{N}\,\text{mm}^{-1}$ at joint *H*. This load distribution is equivalent to a resultant concentrated vertical force of 45 kN acting between joints *F* and *H* at a distance 0.9 m from joint *F*. The horizontal component can be considered to

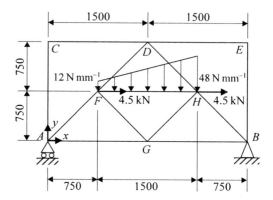

Figure 8.14 Loading on one plane frame section.

consist of two equal concentrated forces of 4.5 kN that act at joints F and H. The loading from the machine is considered to be transferred at the joints into the frame as shear forces only. For this situation to be realized the connections between the machine and the frame have to be pinned, ensuring that there are no moments.

To perform the fatigue design of the welded joints it is necessary to conduct four analysis types of increasing complexity:

1. a pin-jointed frame analysis to determine the axial forces in the members and so to allow the appropriate sections to be chosen from a range of commercially available welded tubes
2. a rigid frame analysis to determine the overall response of the structure and to check the suitability of each of the members in terms of peak stresses and deflections
3. a local analysis of the critical joint H, using a thin-shell finite element model, to obtain estimates of the peak stresses at the welded connections
4. a more refined local analysis of joint H, using a solid finite element model, to obtain more accurate peak stresses in the vicinity of the welds.

Note that analyses 1 and 2 use frame models where the members are modelled as one-dimensional members placed along their longitudinal centroidal axis. These analysis types are often considered as forerunners of the finite element method (Coates *et al.*, 1988). All four analyses can be performed using the finite element method but in this case results will be presented only for types 2 and 3. Also note that the stress results from analyses of types 3 and 4 are used to design the welds against fatigue failure, but there is not the scope in this text to detail the procedure.

8.2.3   Analysis of Pin-jointed Frame

Prior to conducting a finite element analysis to determine the displacements and stresses it is necessary to choose the steel sections to fabricate the frame. To do this the designer needs to know the forces in the members and the simplest technique is to model the frame as a pin-jointed structure. Note, however, that for the frame shown in Fig. 8.14, the distributed loading along member FH induces bending in that member which is transmitted throughout the frame by the joints. This could be important when choosing the section sizes, but will not be predicted by the pin-joint analysis as only axial member forces are determined.

Figure 8.15 shows the plane frame where it is assumed that all joints are pinned, that all members are struts, i.e. they can transmit only axial forces and that, at each joint, the centrelines of the members are coplanar and concurrent. For this analysis the vertical loading is been modelled as acting only at ends of member FH, with the values of 18 and 27 kN determined from an application of static equilibrium to the distributed loading. It is assumed that the loading has a line of action coinciding with the centroid plane of the frame. In practice, there will be eccentricity causing a moment about the x-axis. Our ideal assumption

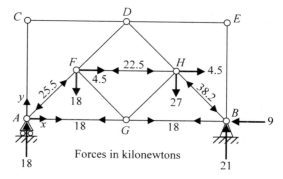

Figure 8.15 Analysis of the plane frame with pin joints.

simplifies the analysis and is valid provided that the resultant out-of-plane moment is small. This is yet another situation where engineering judgement is used to reduce the complexity of the model. If the effect of the out-of-plane moments is to be included, then the model needs to be three-dimensional.

Next, the supports for the frame are modelled such that the whole problem is statically determinate and the reaction forces can be determined from considerations of static equilibrium alone. Joint B is assumed to be fixed in space and joint A is assumed to be free to move in only the horizontal direction, hence joint A is shown in Fig. 8.15 fixed to a roller support.

Some discussion of the choice of these ideal support conditions is necessary. In practice, the frame may be supported by the strong floor of a factory via elastomeric bearings placed at its four corners. These bearings are used to reduce the transmission of the dynamic excitation through the floor into the fabric of the building and, commonly, they are produced from viscoelastic rubber. To work effectively, the bearings must have a stiffness much smaller than that of the frame and its members, but this leads to problems if the bearings are modelled as spring elements in the full finite element analysis as a result of ill-conditioning, as described by Cook *et al.* (1989). To prevent occurrence of this problem, which is not easily remedied, ideal rigid support conditions are used.

By considering the static equilibrium of each joint in turn (Coates *et al.*, 1988), the forces in the members can be determined and these are shown in Fig. 8.15. From these forces the choice of section can be made. Here, sections from the British Steel COLDFORM range of hollow rectangular and square sections are used. To aid the designer, technical data sheet TD 341/20E/91 (available from British Steel's advisory service) provides data for each section, giving an axial tension and compression resistance to prevent buckling. Note that the grade of steel is 50/45 having, for the purposes of structural design, a yield strength (σ_Y) of 350 N mm^{-2} and an ultimate tensile strength (σ_U) of 430 N mm^{-2}. Also, the isotropic homogeneous material has the following elastic constants: Young's modulus, 205 N mm^{-2} and Poisson's ratio, 0.3.

As all members in the frame have to resist a combination of stresses due to axial forces, shear and bending it was decided to choose section sizes that have an axial resistance at least three times the axial force given in Fig. 8.15. Members *AF*, *FD*, *FH*, *DH* and *HB* are thus fabricated from rectangular section material which is 80 mm deep by 40 mm wide and has a wall thickness of 4 mm. All other members are made from square section material 40 mm deep and 40 mm wide, with a wall thickness of 4 mm. This smaller section material is also to be used for the members that connect the two plane frames shown in Fig. 8.13. Note that the deeper section has a tension capacity of 294 kN and a maximum axial compression resistance of 135 kN (minor axis buckling), while the values for the shallower section are 182 and 72 kN, (major and minor axis buckling respectively).

When assembling the frame, the members are connected such that the centroidal axes are coplanar and concurrent, as was assumed for the pin-joint analysis, and so there is no joint eccentricity to make subsequent analyses more difficult. Note also that using just two section sizes, each of width 40 mm, ensures that the fabrication process is straightforward and that the joints have the full section strength. Finally, the actual dimensions of the frame are now 3040 × 1540 × 1540 mm.

8.2.4 Rigid-frame Finite Element Model

If the frame is to be loaded solely at the joints and has a construction such that it is statically determinate then the pin-jointed analysis can be used to determine accurately the joint strengths and deflections (Packer *et al.* 1992). However, this frame has to transmit some bending due to the distributed loading on member *FH*. Hence, to determine a statically admissible set of moments, shear and axial forces throughout the frame, a finite element beam analysis of the rigid-jointed frame can be made.

In Sec. 3.8.1 a linear bending beam element was developed that did not model the effect of shear deformation. Such an element models the prismatic member subject to a loading normal to the centroidal axis by lumping all the properties along this axis and assuming that the bending deformation causes the normals to this axis to rotate but remain normal. These elements have two nodes, each possessing two degrees of freedom and, in standard texts, such an element is referred to as the beam element. Commercial packages generally do not include such a simple two-dimensional beam element alone, rather they include a three-dimensional beam element that possesses six degrees of freedom per node. This element is the combination of four simple elements: a torsional element, an axial element, a major-axis bending element and a minor-axis bending element.

These beam elements are defined within the pre-processor, usually in a module specifically provided to handle beam elements, by specifying the geometry of the section. Geometrical properties for a section are then calculated and are assigned to the appropriate beam elements within a mesh during the solution phase. Here, two structural sections are used and so two sets of element properties need to be defined.

Next, a mesh of these beam elements has to be created to model the geometry of the plane frame in the x–y plane of a global Cartesian axis system, where the origin has been chosen to be at joint A. For a frame structure it is not necessary to consider meshing areas or volumes as the continuum is modelled by one-dimensional elements. To model this frame, the mesh shown in Fig. 8.16 has been constructed by first defining the positions of the 24 nodes and then connecting them with beam elements to which the correct section properties have been attached. Note that the positions of the nodes in the vicinity of joint H have been chosen carefully, as the calculated nodal forces and moments at these positions are used as boundary conditions for the substructure analysis presented in Sec. 8.2.5.

Then the isotropic material properties of the beam elements can be defined. These are taken from technical datasheet TD 341/20E/91 for British Steel's steel Grade 45/50. Note that the analysis needs only the elastic constants for the steel as the stress results from a beam model cannot be readily used in a failure prediction.

Boundary conditions are then entered to define the loading arrangement and support conditions of the model and these are shown in Fig. 8.16. Here, the support conditions at joints A and B are the same as those for the pin-jointed truss model shown in Fig. 8.15. To enforce these restraints the nodal displacements u and v are zero at joint B and the displacement v is zero at A. With the loading acting in the x–y plane there is no requirement to specify, at each node, that the nodal displacement w, and the rotations θ_y and θ_x are zero. These nodal displacements remain inactive automatically.

The loading defined in Fig. 8.14 is modelled in two separate parts. First, the horizontal concentrated forces at joints F and H are entered as nodal forces of 4.5 kN in the x-direction and zero in the y- and z-directions. Then the vertical distributed loading along the length of member FH, shown in detail in Fig. 8.17 as a distributed load with a linear variation, is allocated to each of the five beam elements representing the member FH. This is usually done using a general beam

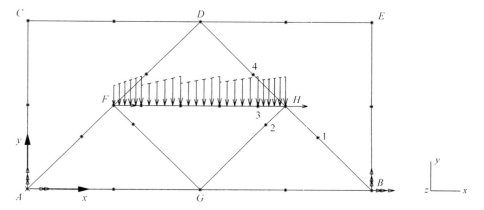

Figure 8.16 A beam element model for a rigid-joint analysis.

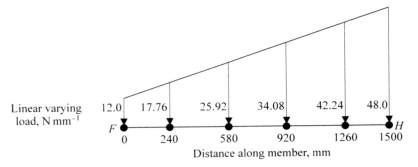

Figure 8.17 The load distribution for the elements along member *FH*.

loading facility where distributed loadings of any variation can be applied. Finally, before the displacements and stresses are calculated, a check is made on the reaction forces to see that they balance the applied external loading.

Figure 8.18 shows the deformed shape of the frame together with the undeformed shape. As anticipated, the largest deflections occur in the horizontal member *FH*. In fact, the maximum deflection is about 5 mm, which is $\frac{1}{300}$ of the span, and this is considered to be within the design specification. By comparing the deformed and undeformed shapes of the frame, it is apparent that joint *H* is experiencing the highest deformations and so is the critical joint in the frame.

Results from the static analysis can also be presented in the form of displays of the beam forces and stresses along the length of the beam elements superimposed on the mesh. Figures 8.19–8.22 show the information for the bending moment (M_z) and axial force (F_a) distributions and their stress distributions. Solutions for the shear force and stress are not given here because shear does not influence decisions in the design of the frame. Note that the bending stresses

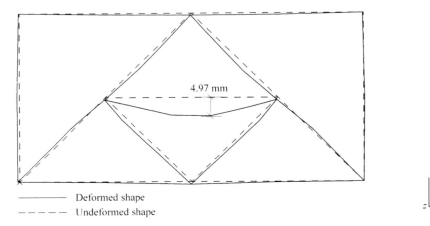

——————— Deformed shape
– – – – – Undeformed shape

Figure 8.18 Displacement of the frame calculated using beam elements.

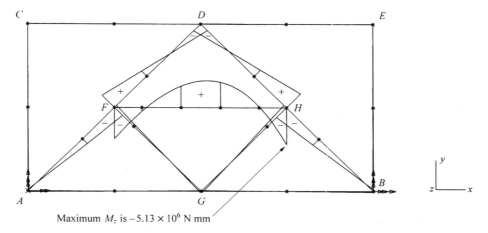

Maximum M_z is -5.13×10^6 N mm

Figure 8.19 Bending moment distribution in the frame.

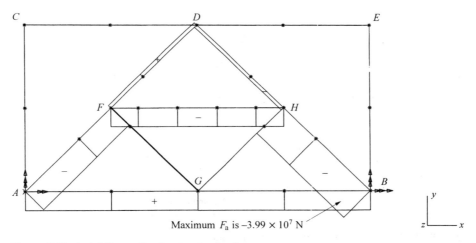

Maximum F_a is -3.99×10^7 N

Figure 8.20 Axial force distribution in the frame.

are calculated from the bending moments using the bending formula of (2.25), and they are calculated as the maximum value, that is at the surface of the section. For example, the maximum direct stress at the mid-span of member *FH* as a result of bending is at a distance of ± 40 mm from the centroidal axis. Clearly, the axial stress in Fig. 8.22 is calculated by dividing the axial force by the cross-sectional area of the beam member, and this geometric property is determined when the section is constructed in the beam creation facility.

Combining the two direct stress components gives the stress output displayed in Fig. 8.23. Note that this combined stress may be given as a magnitude only.

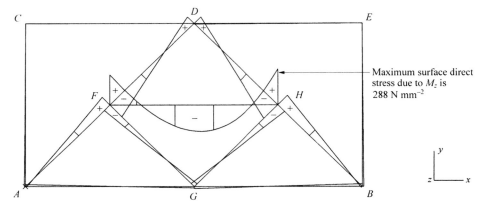

Figure 8.21 Distribution of maximum bending stress in the frame.

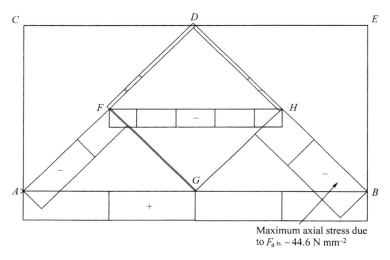

Figure 8.22 Distribution of axial stress in the frame.

This stress can be used to check the suitability of the beam section sizes. Disregarding the areas of the joints, the maximum tensile and compressive stresses are 0.6 of the yield stress given and this is considered to be a safe design.

Focusing on the critical joint, joint H, Fig. 8.24 gives the combined stress values at this joint calculated from the element forces for the four beam elements that meet at this joint. The variation in the values is very high, such that it is impossible to establish the maximum stresses that the butt welds of the joint need to transmit. As it is these stresses that are needed for the fatigue design, they have to be modelled using a substructuring technique. This requires that the boundary

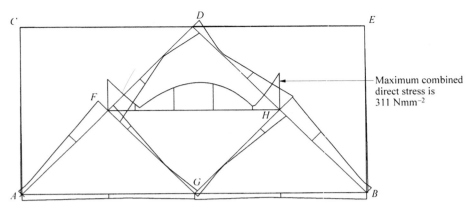

Figure 8.23 Distribution of maximum combined direct stress in the frame.

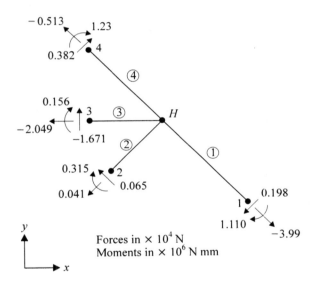

Forces in × 10⁴ N
Moments in × 10⁶ N mm

Nodal displacements				Surface direct stresses at joint H		
Node	u mm	v mm	θ_z mrad	Element	Top, N mm^{-2}	Bottom, N mm^{-2}
1	0.398	0.197	0.92	①	−159	151
2	0.201	−1.158	1.39	②	75	75
3	−0.290	−2.577	8.72	③	265	−311
4	−0.842	−1.175	0.40	④	62	−151

Figure 8.24 Stresses and displacements at joint H from beam model.

conditions are provided by the frame analysis; these are presented in Fig. 8.24 as the forces and displacements at nodes 1 to 4.

8.2.5 Modelling the Substructure with Shell Elements

As was observed above, the stress output from a rigid frame analysis cannot be readily used to design the joints. One approach to obtaining an improved stress solution is to model the hollow sections of the complete frame with thin-shell elements but there are several reason why this is not an attractive option. First, this analysis takes a lot of effort and does not necessarily provide all the information required. Second, it is clear from the frame analysis that joint *H* is the critical joint, and so if it can be demonstrated that this joint design is acceptable in terms of its fatigue performance, then the rest of the butt-welded joints will be acceptable. Finally, as the real purpose of this particular example is to explain how real engineering problems can be modelled, it is more appropriate here to illustrate the important modelling technique known as substructuring.

Occasionally, a finite element analysis will produce a solution which is good over most of the continuum but which lacks accuracy and convergence in one or more local regions. On other occasions, the model may be complex and contain many components such that mesh refinement in all regions becomes impractical, leading to the exhaustion of the computational resources or the limits of the numerical accuracy being exceeded. These common situations call for the ability to model a substructure or subregion alone but subject to the influence of the overall problem under consideration.

There are a few methods available that can be classed as substructure modelling and these are introduced by Knight (1993). Here, an approach will be used that takes the output at specific nodes of the overall frame model and, from this output, establishes the boundary conditions for a substructure model of joint *H*, where thin-shell elements are used to represent the walls of the hollow sections.

Note here that it is expected that a thin-shell finite element model of the joint will give an improved stress prediction in the vicinity of the welds. However, modelling of the welds themselves is largely ignored. For example, the mechanical properties of the weld material are not accounted for, nor is the the actual geometry, as the actual weld thickness is greater than the wall thickness modelled and also fillet radii at the welds are ignored. These factors must be allowed for when interpreting the results of the substructure analysis. If further refinement is necessary to take proper account of the welds, then the solution for the thin-shell element model can be used to provide the boundary conditions for a more localized substructure model of solid elements. Any such solid model will have many elements and degrees of freedom, if element aspect ratios are to be sensible, with one or two elements through the walls of thickness 4 mm being unlikely to give accurate stress predictions. Hence, such an analysis will be very costly both in the time taken to build the model and in the computing resources needed to produce a solution.

Before the mesh can be specified, the geometry of the substructure needs to be defined. As the boundary conditions are generated from the nodal values of the rigid-frame model, the lengths of the members in the region of the joint are pre-determined by the beam element mesh. This is shown in Fig. 8.25, where the finite element mesh is shown superimposed on the local beam elements of the frame analysis. To simplify the behaviour of the joint, and to make both the pinned and rigid-frame models correct, the members have their centre lines concurrent. It is often the case, however, that members are offset such that eccentricity is present.

The shell mesh is constructed within a new global axis system which is chosen with its origin at the point where the centre lines of the four members intercept, i.e. point H. Coordinates for the ends (i.e. nodes 1 to 4) of the substructure are calculated from the frame model, and trigonometry is used to determine the coordinates of the 14 separate areas of the surface where the thin-shell elements will be placed. Simple distributions of elements are then created on these areas of the surface to give the mesh shown in Figs 8.25 and 8.26. Clearly, this is an example of a mapped mesh. Some element distortion has occurred but this is well within the limits imposed by the element formulation. Note, however, that the geometry used here defines the middle surface of the elements and so the depth and breadth of each member are each some 4 mm, i.e. the wall thickness, narrower than the nominal section size.

To assist in fabrication of the physical frame and to improve structural performance, the member joining nodes 1 and 4 is a single continuous length of section $80 \times 40 \times 4$ mm. The other members consist of a length from node 2 of

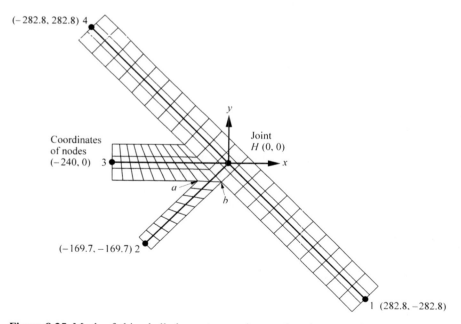

Figure 8.25 Mesh of thin-shell elements superimposed on beam mesh.

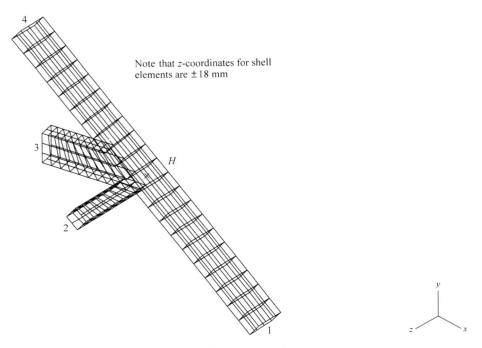

Note that z-coordinates for shell elements are ± 18 mm

Figure 8.26 Isometric view of thin-shell element mesh.

section $40 \times 40 \times 4$ mm and a length from node 3 of section $80 \times 40 \times 4$ mm. In the model the connection between these sections has been approximated as can be seen in Fig. 8.25. Looking at the surface labelled ab, it is found that it is not horizontal and has a downward inclination of $1.9°$. This slight modification to the mesh prevents the model from having elements of widely different aspect ratios. Such a modification to the geometry is not considered serious here, as this region is in compression and it is the tensile regions of the butt welds that limit the fatigue resistance.

The model was constructed from linear quadrilateral thin-shell elements, and there are 374 elements and 384 nodes. Each node has 6 degrees of freedom and so there are 2304 degrees of freedom in total. All elements are assigned a thickness of 4 mm and have the material properties given above. This shell analysis allows von Mises' stress contour plots to be generated and this requires values for the direct yield strength.

8.2.6 Applying Boundary Conditions to the Substructure Model

Now the boundary conditions must be specified. The first of these is a restraint on all movement at one node in the frame model, which translates to support restraints for a plane of nodes. Noting, from the frame model, that the smallest rotation θ_z is at node 4 it is decided to fully restrain the nodes at the equivalent plane of nodes, that is all six degrees of freedom are set to zero. This support boundary condition, shown in Figs 8.27 and 8.28, is not physically correct and any stress concentrations that arise near to this region has to be ignored. However, this is a valid assumption if advantage is taken of Saint-Venant's principle. This tells us that any local behaviour dies away quickly away from a local disturbance such that results are unaffected a reasonable distance from where any slightly erroneous boundary conditions are applied. Saint-Venant's principle also allows us to model, at nodes 1 to 3, the loading due to the element forces, be they axial or shear forces or bending moments, given in Fig. 8.24, as a statically admissible set of edge pressures. The word 'admissible' is important here as it means that the loading in the model does not exactly match the true distribution but is acceptable for the purpose of modelling.

To illustrate the modelling procedure for the loading at the ends of members, the calculation of the constant edge pressure, entered into the software as a force per unit length per element thickness (dividing this force by the element thickness gives the pressure loading in N mm^{-2}), shown in Fig. 8.29 for node 1, will now be considered. Note that in the figure the directions used to define the sign of the forces are shown.

There are three element forces to consider. First, the axial load at node 1 is $-3.99e + 04\,\text{N}$. To determine the axial edge pressure of $-178.2\,\text{N mm}^{-1}$ per

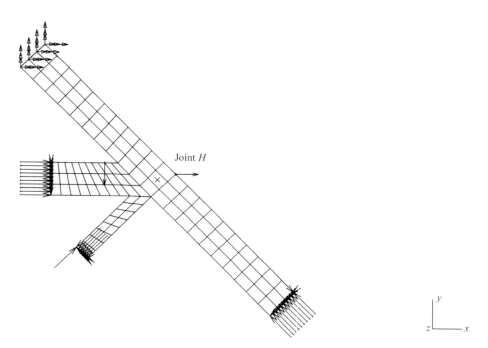

Figure 8.27 Boundary conditions for shell model of substructure.

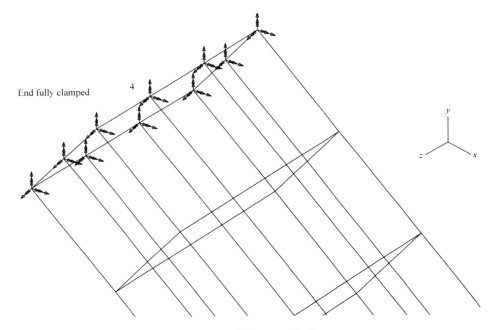

End fully clamped

4

Figure 8.28 Fully clamped boundary condition at node 4.

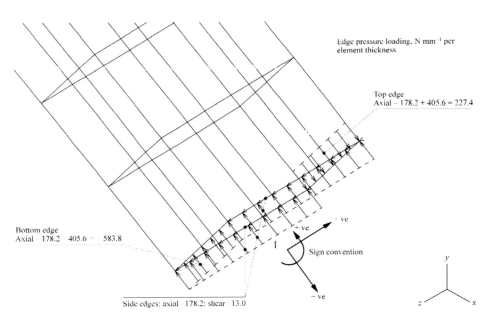

Edge pressure loading, N mm⁻¹ per element thickness

Top edge
Axial – 178.2 + 405.6 = 227.4

Bottom edge
Axial – 178.2 – 405.6 = – 583.8

Side edges: axial –178.2; shear –13.0

Sign convention

Figure 8.29 Boundary conditions applied at node 1.

element thickness, the load is divided by the length over which it acts, 224 mm (twice 76 mm plus twice 36 mm). This axial pressure is applied to all four edges. Second, the shear load of $-1.98e + 03$ N is assumed to act as a constant shear stress of -13.0 N mm^{-1} per element thickness over the two side walls. Finally, the bending moment of $-1.11e + 0.6$ N mm must be modelled. The simplest way is to use the fact that the moment is a couple that may be modelled by two equal and opposite parallel forces separated by a distance, i.e. $M = Fd$. By taking d as 76 mm, the distance between the top and bottom edges of the mesh, the value of the force F is $1.46e + 04$ N. This force is translated into an axial pressure loading of 405.6 N mm^{-1} per element thickness by dividing it by 36 mm, the length of an edge. Following the direction of the moment it is positive at the top edge and negative at bottom edge. The software takes account of the constant wall thickness.

Now that the boundary conditions are known in terms of the data derived from the rigid-frame analysis, this stage of the modelling process can be completed by including the loading due to the external applied loads. This loading consists of two parts:

1. A concentrated force of 4.5 kN acting in the x-direction with its (assumed) line of action being through the centre of joint. In the real frame this load is applied to the side walls (front and back) of the member 1H4, as shown in Fig. 8.25, and, for example, this may be achieved by having a rigid bar member welded to both front and back walls. Our model of the substructure does not include this level of detail and so the load is modelled in a different way. Here, note that it is important to maintain symmetry in the problem and so to have all of the loading acting in the x–y plane. This eliminates any out-of-plane deformations which would make the rigid-frame analysis invalid. A statically equivalent concentrated nodal force and moment are, therefore, applied to a mid-position node on the top surface of the member, and this loading condition is, in part, shown by a horizontal arrow in Fig. 8.27.

2. A distributed vertical loading on member FH (see Fig. 8.17). On studying the geometry of the substructure in Figs 8.25 and 8.26, it can be seen that the length of member 3H is less than the corresponding sections in the frame analyses. This is because of the physical geometry of the members which are now being modelled. Consequently, the distributed load cannot be applied as shown in Fig. 8.17. As an approximation, it has been decided to model this loading as a statically equivalent single vertical force of 10.83 kN at a distance of 122.5 mm from the end of the member, as shown in Fig. 8.27. It was necessary to model this load as a force and compensating moment as the nearest node to this position where the force acts is actually 120.2 mm from the end.

8.2.7 Analysing the Results for the Substructure

Having applied all the necessary boundary conditions a static analysis of this finite element model of the substructure is carried out. As a first check of the results of this analysis, the reaction forces (F_x, F_y and M_z) at the restrained end are checked to ensure that static equilibrium is satisfied and so to establish that the loading is that given by the frame analysis. As this simple check shows that the overall position is correct, a more detailed analysis of the results can be made.

By using post-processing facilities the results of the analysis can be viewed. For example, the deformed shape and stress contour plots of any view of the model may be produced. Here, however, it is important to plan the required pictures so that they are meaningful to those who have to interpret them.

For this analysis let us consider the deformed shape, which is shown in Fig. 8.30 together with the undeformed shape. Note that the displacements are greatly magnified and that the display produced here is a two-dimensional view for the x–y plane and has been chosen to show that the joint is not twisted out of this plane. Comparing the deformed shape calculated from the substructure analysis with that for the same joint calculated from the frame analysis, Fig. 8.18, it is seen that the two analyses predict the same overall behaviour. This positive check between the analyses is another piece of evidence that points to the substructure model being qualitatively correct.

As to the use of the substructure model in the design of the frame, it is the results for stress within the material of the joint that are needed to carry out a fatigue analysis of the butt welds. The designer will want to know the peak stress values when the dynamic loading is a maximum and a minimum and the peak

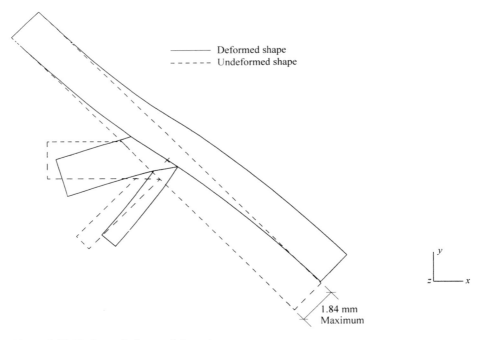

Figure 8.30 Deformed shape of the substructure.

stresses when the machine is not in motion and the frame has to transmit only the dead loads. The finite element model developed above will predict the first of these situations and so further analyses are necessary to predict the other cases.

To illustrate a typical display of stress results, consider the the butt weld in joint H with the highest tensile stresses. By generating a stress contour plot of the maximum principal stresses, e.g. σ_{ts} at the top surface of the shell elements (see Fig. 7.22), over the whole model and then using the zoom facility, it is a simple task to find that this critical connection occurs where the top surface of the horizontal member $3H$ meets the bottom surface of member $4H$.

To present the contour displays, two areas of elements have been chosen and the elements grouped for display purposes. Figure 8.31 shows that such a display effectively isolates the required butt weld from rest of the mesh. Clarity of the stress contour display is improved by magnifying the region of the weld on the screen and then displaying only the boundaries of the element group. Figures 8.32–8.34 show, respectively, contour plots for:

- the maximum principal stress at the top surface σ_{ts}, which comprises mainly tensile stresses
- the maximum principal stress at the middle surface σ_{ms}, which is mainly axial stress
- the maximum principal bending stress σ_B.

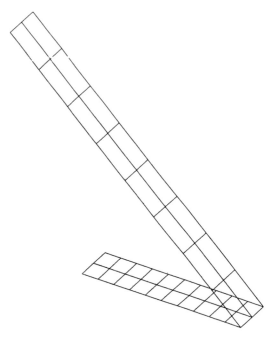

Figure 8.31 Group of elements for display purposes.

The contours are generated using nodal values from all nodes in the model and not just those nodes associated with the elements shown in Fig. 8.31. If the plots are generated using only those elements in the group shown, then the values displayed are restricted in range, producing a misleading impression. Hence it can be seen that an analyst has to ensure that the contour plots are calculated using all relevant nodal values to give a meaningful display.

Figures 8.32–8.34 also show three nodal stress values along the line representing the welded connection. From this the peak stress value is seen to be about $150\,\mathrm{N\,mm^{-2}}$. Comparing this with the value calculated using the rigid-frame analysis, i.e. $265\,\mathrm{N\,mm^{-2}}$ shown in Fig. 8.24, and with the yield stress of $350\,\mathrm{N\,mm^{-2}}$, it can be seen that the shell model produces a peak stress at the weld about one-half of the beam analysis value. Figures 8.33 and 8.34 show that the top surface stress is principally due to an axial component with the bending component being significantly lower. Note that the relationships between the stresses given in Fig. 7.22 do not hold at the location of the butt weld, because not all of the elements are in one plane.

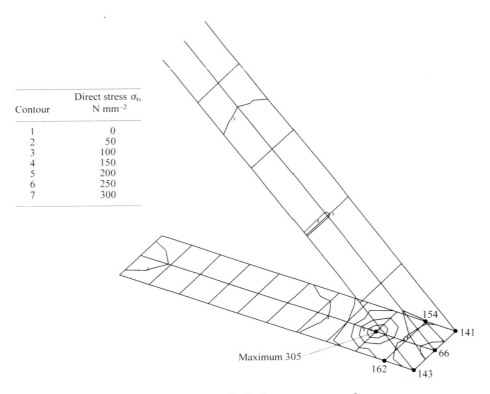

Contour	Direct stress σ_{ts} N mm^{-2}
1	0
2	50
3	100
4	150
5	200
6	250
7	300

Figure 8.32 Contour plot of maximum principal stress at top surface.

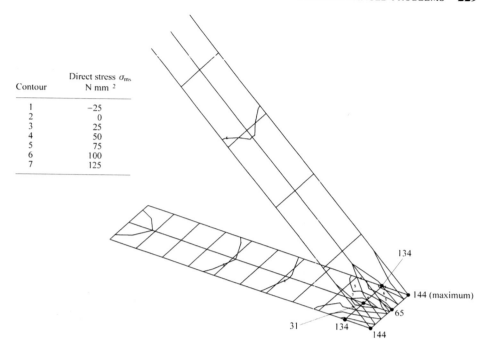

Contour	Direct stress σ_{ms} N mm^{-2}
1	−25
2	0
3	25
4	50
5	75
6	100
7	125

Figure 8.33 Contour plot of maximum principal stress at middle surface.

Finally, Fig. 8.35 is a contour plot of the von Mises' maximum principal stress at the bottom surface of the shell elements, the location where this stress is maximum. This is derived from the standard formula

$$\sigma_{\text{vonMises}} = \sqrt{\frac{1}{2}[(\sigma_1 - \sigma_2)^2 + (\sigma_1 - \sigma_3)^2 + (\sigma_2 - \sigma_3)^2]} \qquad (8.8)$$

where σ_1 and so on are the principal stresses. The formula predicts that yielding occurs when the value of von Mises' stress in (8.8) exceeds the yield stress in uniaxial tension σ_Y which is 350 N mm^{-2} for grade 45/50 steel. It can be concluded that yielding of the material has not occurred in the region of the weld because the peak von Mises' stress there is about 150 N mm^{-2}.

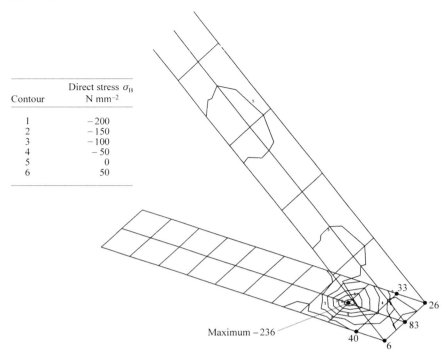

Contour	Direct stress σ_B N mm^{-2}
1	−200
2	−150
3	−100
4	−50
5	0
6	50

Maximum −236

Figure 8.34 Contour plot of maximum principal bending stress.

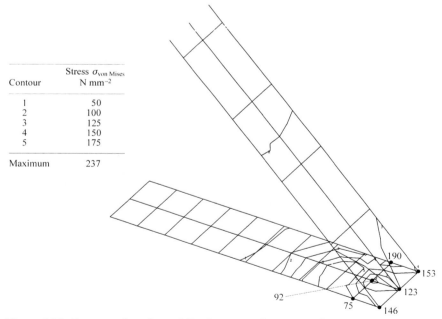

Contour	Stress $\sigma_{\text{von Mises}}$ N mm^{-2}
1	50
2	100
3	125
4	150
5	175
Maximum	237

Figure 8.35 Contour plot of von Mises' stress at bottom surface.

9

INDUSTRIAL CASE STUDIES

Having discussed four examples in detail within Chapters 7 and 8, where minimal computing effort was required, some larger cases can now be considered. These have an automotive flavour, but they are typical of the larger scale studies that are carried out in a variety of industries. While the previous examples enabled discussion of some of the finer points of modelling, these industrial case studies will show how finite element modelling is used for situations with complex geometries where some form of assumption has to be made to progress a solution. It is the use of sensible assumptions in given situations that distinguishes the artistry of truly professional analysts.

The following cases will be studied here:

- a strength and stiffness analysis of the frame of a recumbent bicycle
- an analysis of a connecting rod from a road vehicle engine
- a composite suspension arm from a road vehicle
- a displacement analysis of a passenger vehicle body shell.

9.1 ANALYSING A BICYCLE FRAME

In this example the strength of a commercially available recumbent bicycle which has been used as the basis for a variety of undergraduate projects in the area of human-powered vehicles is considered. To define the geometry, a solid model is built and then the surface of this is meshed with thin-shell elements. Realistic load conditions can then be simulated and the strength of the existing frame assessed.

9.1.1 Description of the Problem

The generic form of a recumbent bicycle is illustrated in Fig. 9.1, which is a solid model of the major components. It is clear from this figure that the rider sits in a quite unusual position, compared to that for a normal bicycle. In this case the rider sits as if in an armchair and provides motive power by pushing on the pedals which are located directly ahead. With the rider in this position, the aerodynamic drag of the rider–bike combination is reduced and this enables higher speeds to be achieved for a given power input. For this reason, such a riding position has been widely adopted by several manufacturers of human-powered vehicles. As part of a programme of undergraduate project work, ways of improving the design in terms of the aerodynamics, steering, ergonomics and drivetrain of the vehicle have been investigated, together with a study of putting a small engine on the bike to turn it into a low-powered endurance vehicle.

Before making any structural changes to the bike, the strength of the frame needs to be assessed and any weak areas determined. This enables strengthening to be carried out where necessary before any modifications are made that might increase the load on the bicycle frame. Stress results will be produced for this geometry under normal static load conditions.

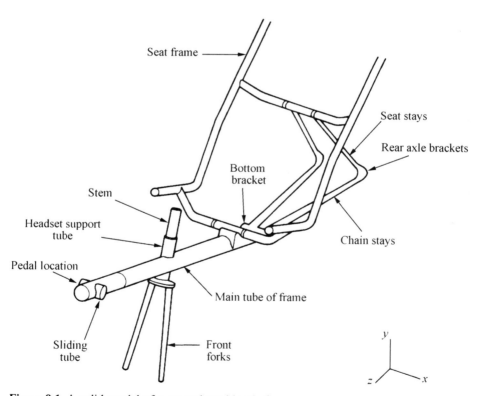

Figure 9.1 A solid model of a recumbent bicycle frame.

9.1.2 The Bicycle Geometry

A detailed study of Fig. 9.1 shows that this bicycle is of extremely simple construction. A conventional bike has a triangular frame with the main triangular section made of a top tube, seat tube and down tube. In the recumbent design studied here, this triangle of tubes is replaced by a single structural member which is a tube of larger diameter than is used on a conventional bicycle. Through this main member is placed a tube which supports the bearings, or headset, for the stem attached to the handlebars and the front forks. For simplicity of manufacture, these two tubes are welded at right angles to each other. Elliptical cross-section forks are used which are straight and welded to a cross-piece, the fork crown. To support the pedals and chainwheel, a smaller tube which holds an axle and bearings is slipped into the front of the larger tube and clamped in place. At the rear of the main tube a conventional, but nonfunctional, bottom bracket supports two chain stays, to which the rear axle and the gear system are clamped. Finally, two seat stays are joined to the top of the tubular seat frame and a bracket clamps the seat frame to the main tube. Note that webbing is slung across the seat frame to support the rider.

Detailed measurements of all of the relevant structural members have been taken manually from the physical frame and from this data the solid model shown in Fig. 9.1 can be created. Note that all the members are of single thickness as no joints have been double-butted.

9.1.3 Creating a Solid Model of the Geometry

Care must be taken when thinking about the creation of the solid model. Yet again, it is seen that the needs of the analysis must be taken into account before producing a computer model. In this case a thin-shell finite element model will ultimately be built; this dictates the approach to be taken in creating the solid model of the geometry. It is perfectly feasible, and probably very obvious, to build a solid model of the exact geometry using tubes for all the components that are tubular. Here, however, as shell elements are to be used which have zero physical thickness, a different approach has to be taken. The mathematical formulation for an element handles the behaviour of the element owing to its thickness and so the elements must be placed with their nodes at the mid-depth position in the out-of-plane direction. This is the same procedure as that used for the pressure vessel in Sec. 7.2 and for the substructure analysis of the frame joint in Sec. 8.2. To ensure that this is easy to do, the physical dimensions of the tubes are not recreated in the solid model, rather the tubes are modelled as solid cylinders with a diameter calculated to allow for the actual material thickness.

First, the main tube is modelled using the primitive shape of a cylinder and is placed in the correct orientation. Then further cylinders to represent the holder for the forks and stem, and the bottom bracket, are created and placed in position. Using the Boolean addition operation typical of solid modelling software that uses constructive solid geometry (CSG), these smaller cylinders can then be

joined to the main cylinder. The chain stays are created using a skinning operation. To do this closed curves, known as profiles, of the appropriate form for the
ends of the stays are created and a smooth skin placed between them. The software uses a smooth surface to transform these profiles into solid objects. These
stays can then be placed in position and added to the cylinders. To complete the
first stage of the geometry modelling, the rear axle brackets are created as a profile
in two dimensions and then extruded through the third dimension to give the
object depth, before being joined to the rest of the model. Figure 9.2 shows the
solid model as a wireframe form at this stage of the process.

   Next, skinning operations are used to model the seat stays. Also simple clips
and brackets are modelled to support the seat frame, before the two cylinders
representing the pedal support are joined and moved into place. Figure 9.3 shows
the model at this stage.

   A further skinning operation is used to form, the seat frame, with a circular
profile being swept along a path in three dimensions for each of the four members
of the seat frame. In those areas where the seat frame is highly curved, a number
of profiles are used to ensure that a smooth surface is created. Finally, skinning is
used to create the front forks which are attached to the stem through a cross-
piece, the fork crown. With these items added to the model the necessary structure
is complete and a wireframe of the complete model is shown in Fig. 9.4.

9.1.4   Obtaining a Mesh of the Geometry

To produce a mesh from the solid model of the geometry an automatic free mesh
generator is used. This looks at the solid model, then places a pseudo-triangular

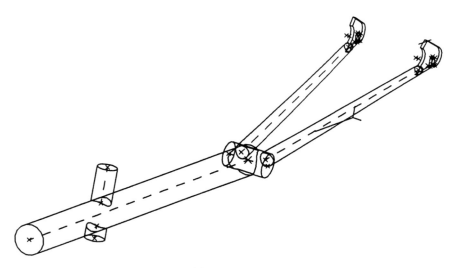

Figure 9.2 First stage in the assembly process.

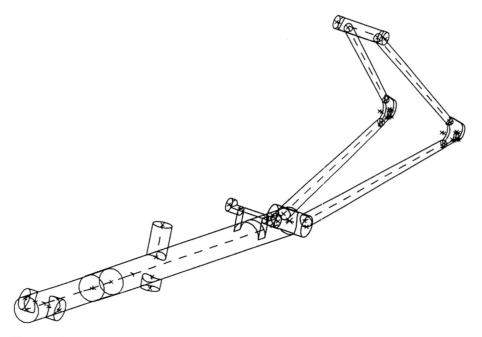

Figure 9.3 Second stage in the assembly process.

mesh on the surfaces of the model before creating a final tetrahedral mesh within the surfaces. In this case the tetrahedral mesh is deleted, as it is not needed for this thin-shell finite element analysis. However, the data defining the surface patches enclosing the volume is retained, so that a thin-shell mesh can be created on these surface patches.

Separate meshes have been created for the forks–stem assembly, the seat frame, the pedal support tubes and the main structure. To control the distribution of the elements, global element lengths of 30 mm on the main tube and 10 mm in other areas have been used. A second mesh control parameter has also been used which ensures that a suitable number of elements are placed on highly curved surfaces. Figure 9.5 shows how this works in the two-dimensional case, where the deviation is defined for a single element as a percentage of the element length. The shorter the element, the more accurate is the description of the surface and the lower the deviation. Here the maximum allowable deviation has been set to 20 per cent.

Figure 9.6 shows the full thin-shell mesh for this problem. Note that, for now, only tubular material has been meshed and the solid metal parts, such as the axle brackets, have been left out of the finite element model. Such solid material is considered to be lightly stressed, with the large stresses of interest being generated only in the thin tubing. For this situation, some 15 000 linear triangular elements have been used together with some 8000 nodes.

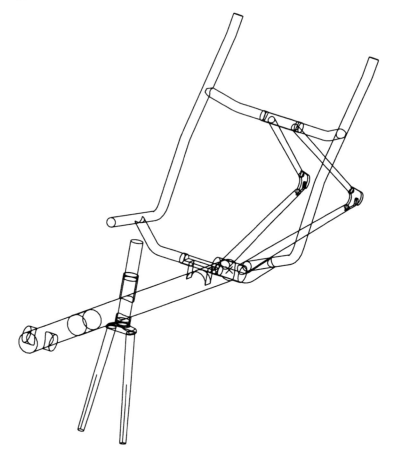

Figure 9.4 Final stage in the assembly process.

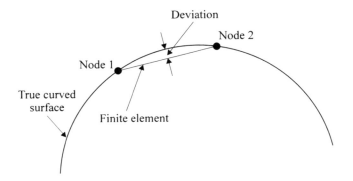

Figure 9.5 Deviation of an element from a surface.

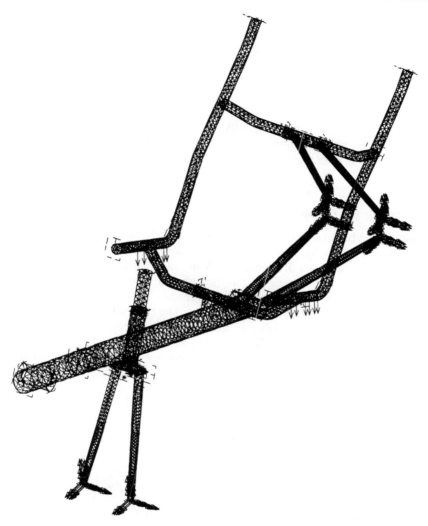

Figure 9.6 The completed finite element mesh.

9.1.5 Material Properties and Application of Boundary Conditions

All of the tubes modelled are of structural steel, as is the seat frame, and appropriate Young's modulus and Poisson's ratio have been defined.

For the strength analysis of the bike only the static load of a rider sitting on the seat will be simulated, i.e. no load is to be applied to the pedals. To do this the nodes that are next to the missing rear axle brackets are restrained such that they have zero displacement in all directions, and the nodes where the front axle is clamped are restrained to have no vertical or sideways movement, from the rider's viewpoint, but they are left free to move forwards and backwards.

The dead weight of the rider is applied by splitting a total load equivalent to 80 kg across 20 nodes along the seat frame. This is shown in Fig. 9.6 by the single-headed arrows, together with the other restraints. Such an approximation is considered to be valid as the area of concern is the bike frame itself not the seat frame. Hence any load applied to the seat should be diffused through the seat frame, ensuring that the load transfer from the seat into the main structure is realistic.

Finally, the seat frame is connected to the clips by rigid bar elements. These are also used to join the pedal support to the main tube and to join the forks–stem assembly to the main tube system.

9.1.6 Analysing the Results

Once the model has been run, the results can be analysed. To do this the graphical post-processor is used, as the large quantity of data produced prevents the raw information being used in the first instance. A useful starting point is to analyse the results for displacement. Figure 9.7 shows a typical view of the structure with a displaced side elevation. Note that the displacements have been magnified and that they show the basic form of the deformation. The frame acts by the load on the seat deforming the main member in a downwards direction in the middle forcing the front end to rise up. This bending is opposed by bending in the forks. Also, the chain stays and seat stays bend to oppose the bending of the seat frame and main member.

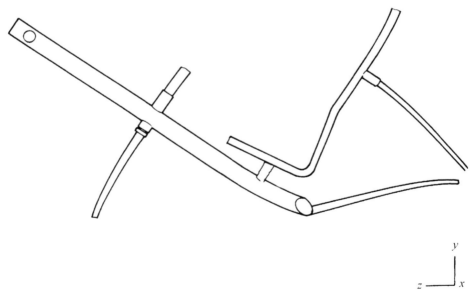

Figure 9.7 Displacement of the frame.

For quantitative data about the behaviour of the structure, contour plots of stress magnitude are shown in Figs 9.8 and 9.9, where the von Mises' stress for certain areas of the mesh are plotted. In reality colour plots would be used but, for reasons of clarity in monochrome, labelled contours are shown here. In Fig. 9.8, the von Mises' stress on the bottom surface of the shell elements are shown in the region of the seat, main tube and stays. Similar plots are obtained for the stresses at the top and middle surfaces of these elements as well. From this figure stresses at contour level 4 (200 MPa) can be seen to occur where the main tube meets the chain stays and where the seat is attached to the main tube. In Fig. 9.9 a close-up of the area around the tube–stay junction is shown where contours of level 6 (300 MPa) occur. As typical values for the ultimate tensile strength of steel vary from 400 to 700 MPa, the finite element model seems to be predicting that failure is close in this case. This analysis shows that strengthening needs to be considered in this area of the structure.

9.2 DESIGN OF A CONNECTING ROD

All piston engines use a connecting rod (con-rod) to convert the reciprocating motion of a piston to the rotary motion of a crankshaft. Traditionally, con-rods have been made of steel, but with the pressure on vehicle manufacturers to reduce the mass of vehicles and so make them more fuel efficient, other lighter materials are now being used.

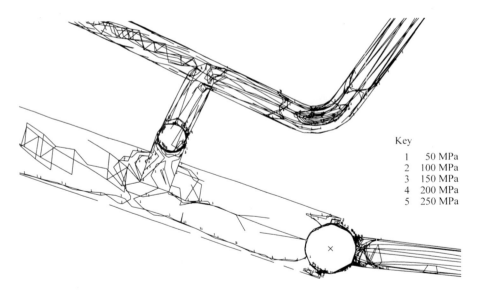

Key

1	50 MPa
2	100 MPa
3	150 MPa
4	200 MPa
5	250 MPa

Figure 9.8 Contour plot for stress—side view.

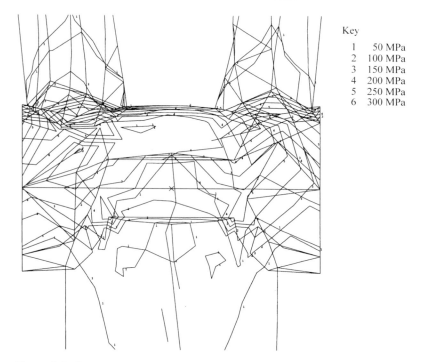

Key

1	50 MPa
2	100 MPa
3	150 MPa
4	200 MPa
5	250 MPa
6	300 MPa

Figure 9.9 Contour plot for stress where stays meet main tube.

This example has been provided by Jaguar Cars and is concerned with the early development of an experimental aluminium alloy con-rod during the late 1980s. During fatigue testing of an initial prototype design, the con-rods were failing well before their specified design life. Investigations were carried out using finite element analysis with the goals of understanding the reasons for these failures, of improving the design before further testing was carried out and of producing appropriate design rules.

9.2.1 The Problem Statement

Finite element analysis is to be used to investigate the fatigue life of a prototype aluminium alloy connecting rod. Modifications to the design can also be considered to ensure that the fatigue life is acceptable. By using such a computational tool near the start of the design cycle, the amount of physical testing required to achieve a successful design should be reduced.

9.2.2 The Geometry of the Con-rod

From engineering drawings of the con-rod a simple solid model can be built, as shown in Fig. 9.10. This shows the shank and little end of the rod together with

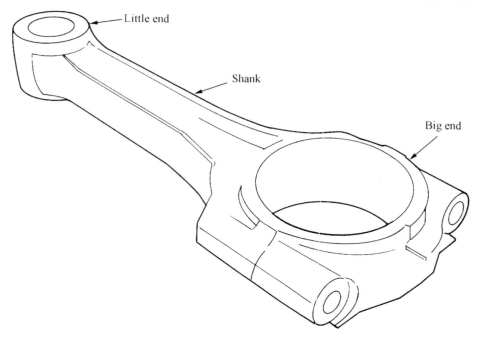

Figure 9.10   Solid model of the connecting rod.

the big-end cap modelled as a single piece, although in fact they are split at right angles to the plane of symmetry of the con-rod along the axis of the crankshaft. Note the holes for the gudgeon pin at the little end, for the crankshaft at the big-end and for two bolts (at the bottom right of the figure). Also extensive use of filleting has been made.

9.2.3   Producing a Mesh

For this design, all the loading cases will be along the longitudinal axis of the rod, and so from symmetry only one-quarter of the physical artefact need be modelled. Hence only one-quarter of the solid model needs to be transferred between the solid modelling program and the finite element pre-processor. Simple mapped meshes can be built to represent the little end and shank of the rod together with the small-end bearing bush and the gudgeon pin. In the big-end region, the geometry is too complex for a mapped mesh and so a mesh has been created manually from the geometry of the solid model. Mapped meshes can, however, be used for the big-end shells and the con-rod bolts, with a manual mesh being used for the crankshaft to match the mesh on the con-rod and cap.

Throughout the model 20-noded parabolic solid elements have been used. Figures 9.11 and 9.12 show the completed mesh with the latter figure showing detail near the bolt. Note that this model was produced before the widespread

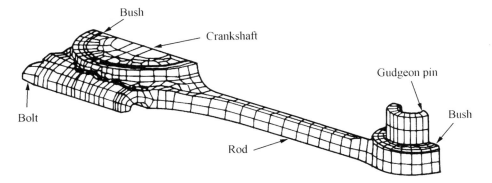

Figure 9.11 Mesh of the connecting rod.

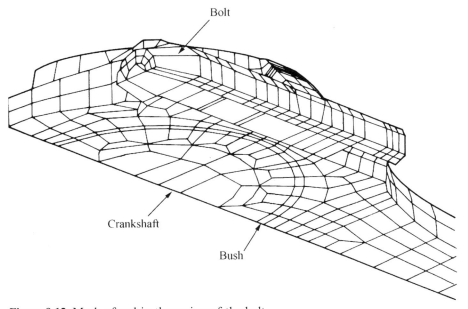

Figure 9.12 Mesh of rod in the region of the bolt.

availability of automatic meshing procedures. Meshing this combination of objects is much simpler now, as was seen with the bike example in Sec. 9.1.

9.2.4 Material Properties and Application of Boundary Conditions

Aluminium alloy is used for the con-rod with steel for the bolt, gudgeon pin, crankshaft and bearings. Appropriate material properties have been used for the elements in these areas.

As the connecting rod assembly is made up of several parts, complex nonlinear interactions govern how these parts transmit forces. Solving the full nonlinear model is prohibitively expensive, and so a series of well-proven engineering assumptions are used to model the interactions while still using only a static analysis. In all cases friction forces are assumed to be zero where metal-to-metal contact occurs. Hence the big-end cap is joined to the con-rod by axial ties which allow lateral movement along the plane of the joint. Similarly, bearings are tied in the radial direction only.

At the bolt the thread is simulated by tying the bolt to the con-rod and cap in the axial direction and the bolt head is tied axially to the cap. Also the bolt is coupled laterally at two points in the thread region and at the head to allow bending loads to be transmitted to the bolt. To simulate the pre-load on the bolt which draws the con-rod and cap together, the bolt is split into three sections with constraint equations used to pull the sections together and so load the whole structure. This load was adjusted until the stress in the bolt after solution was correct. Hence several solutions had to be run to achieve this initial situation. The temperature effects used to do this will be mentioned in Chapter 10.

Contact areas between the gudgeon pin and small-end bush and between the crankshaft and big-end shells are modelled as a contact patch of constant area some 60° to each side of the loading line. This is done with constraint equations, which are different for the contact patches for the different loading cases. Further, interference fits are used to pre-tension the physical bearings at both the big and little ends. This has been simulated using temperature effects again to produce an initial size for the bearings consistent with the interference.

Symmetry conditions have also been applied along the symmetry planes as was explained for the pressure vessel in Sec. 7.2.

The following three simulations have been run with this model:

1. An assembly case where only the effects of the interference fit of the bearings and of the bolt clamping load are simulated. A tightening torque of 50 Nm giving an axial load of 33.3 kN was applied by trial and error, as mentioned previously.
2. A tensile load applied at the crank-pin centre line with all the nodes along this line being constrained to move together. The gudgeon pin was restrained here to provide the necessary reaction to the applied load. This simulates the physical fatigue test not the actual running condition of the engine. The maximum tensile load during the fatigue test was estimated as 15.71 kN and so an assumed load of 20 kN has been simulated by applying a load of 5 kN to the model of one-quarter of the geometry.
3. A compressive load applied at the crank-pin centre line to give the opposite loading situation to that above.

9.2.5   Some Results

Plots of the displacement for the three test cases of assembly, compression and tension are shown in Figs 9.13–9.15, respectively. The displacements have been

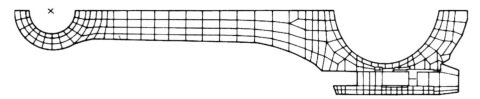

Figure 9.13 Displacement of the model for assembly loading.

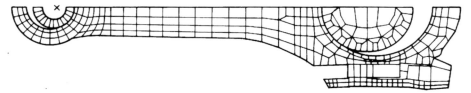

Figure 9.14 Displacement of the model for compressive loading.

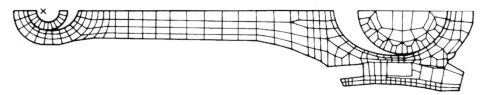

Figure 9.15 Displacement of the model for tensile loading.

magnified by 50 times in all cases. Note that in the assembly case some distortion of the cap and the bolt can be seen. Under compression the little-end eye distorts, as do the cap and bolt yet again, while under tension the eye is stretched severely and the cap–bolt combination distort even more.

Von Mises' stress is shown in the contour plots in Figs 9.16–9.18. The vulnerable areas, i.e. the small-end eye and the cap near the bolt hole can be identified. Note that the compressive load does not alter stresses much from those with just the assembly load at the cap, but the tensile load makes the stresses much larger in this area. Also, the stresses in the eye area are large for both compressive and tensile cases.

Given the results from these three stressing cases, a fatigue analysis can be carried out as the alternating stresses and mean stresses are available for the whole geometry. Such a fatigue analysis predicts a low fatigue life of one million cycles compared to a design expectancy of one hundred million cycles. By increasing the thickness of the material within the rod and repeating the analysis, all the vulnerable areas can be eliminated and the fatigue life improved. Also, design rules for future designs can be derived from these analyses, enabling engineers to design aluminium alloy con-rods in future with the benefit of this piece of work.

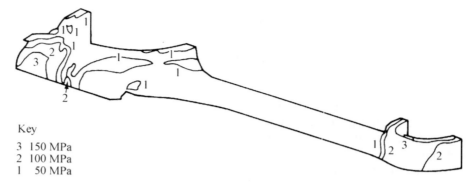

Key

3 150 MPa
2 100 MPa
1   50 MPa

Figure 9.16 Von Mises' stress from assembly loading.

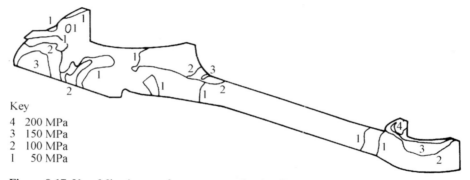

Key

4 200 MPa
3 150 MPa
2 100 MPa
1   50 MPa

Figure 9.17 Von Mises' stress from compressive loading.

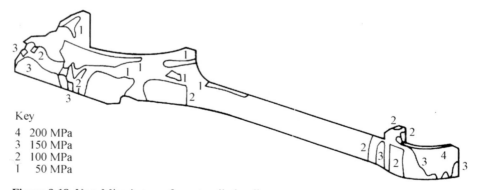

Key

4 200 MPa
3 150 MPa
2 100 MPa
1   50 MPa

Figure 9.18 Von Mises' stress from tensile loading.

9.3 A COMPOSITE SUSPENSION ARM

9.3.1 Objectives of the Simulation

There is a growing interest across many fields of engineering in the use of new materials, such as composites, as a replacement for traditional steels. The driving force is the intention to develop components that are lighter than those made with traditional materials. As with all design problems, some means of assessing the suitability of a component has to be made, and in this case finite element analysis may be able to predict the behaviour of components before the testing of prototypes takes place. If this can be proven, then such analysis can be used routinely in the design process.

As an example of the analysis of a composite material, a finite element analysis of the suspension arm shown in Fig. 9.19 will be described. This is part of a large program where the results derived from computational methods have been compared with the results gained by a variety of experimental methods. The arm is attached to the vehicle subframe through rubber bushes placed at the body mounts shown in Fig. 9.19, with the load applied at the ball-joint housing. The material used for the arm is sheet moulding compound (SMC). This numerical experiment has been summarized by Pinfold and Calvert (1994).

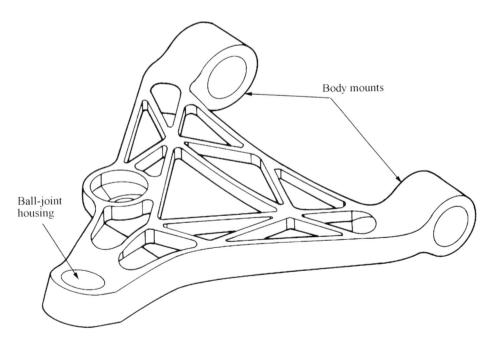

Body mounts

Ball-joint
housing

Figure 9.19 The suspension arm.

9.3.2   Building the Finite Element Model

The geometry of the arm was initially defined within a CAD system as a wire-frame model. To build the mesh shown in Fig. 9.20, sections of the wireframe model were taken such that, where possible, regions of the mesh could have a regular structure. This is clearly the case in the regions of the body mounts and the webs of the structure. Owing to the complexity of the geometry, this strategy could not be applied everywhere and so these regular blocks of elements were joined together manually to form the final mesh. Hence the finished mesh has 1267 solid linear elements of which the majority are bricks, but with some wedge and tetrahedral elements. This is, of course, a relatively coarse mesh.

Compression moulding was used to manufacture the component and despite this the material properties were assumed to be isotropic. Further, from

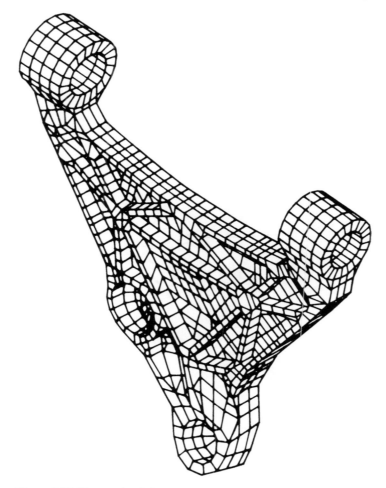

Figure 9.20 The mesh of the arm.

laboratory tests, the SMC has a Young's modulus of 10.5 GPa, a Poisson's ratio of 0.26 and a density of 1800 kg m^{-3}.

To provide loading boundary conditions, the design case known as the *pot-hole brake* condition is used. This simulates a vehicle travelling at 30 mph, where one wheel falls into a pot-hole with the brakes fully applied. From this situation and the vehicle and suspension geometry, the resultant fore and aft loads and also lateral loads on the arm at the ball-joint housing can be calculated. These are applied at the nodes near the housing, with restraints being applied at the body mounts. Note, however, that spring elements have been added to simulate the rubber bushes and that the forces have been applied at the ball joint using beam elements.

Two other load cases were also run to simulate the experimental tests carried out on the component. These experiments were carried out at a reduced load of approximately one-quarter of those at the design condition.

9.3.3   Some Results and Comparison with Experiment

Figure 9.21 shows the predicted stress distribution on the arm. Areas of high stress can be seen in the ball-joint region and near the upper right-hand body mount. These values can also be compared to the results determined through strain gauges placed on the physical arm and through photoelastic techniques. Table 9.1 details

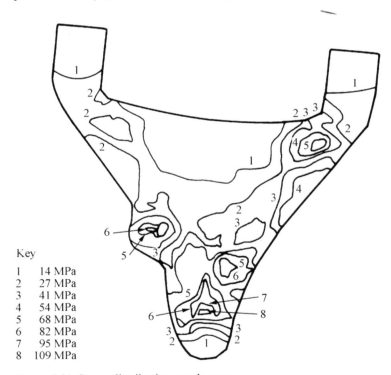

Key

1	14 MPa
2	27 MPa
3	41 MPa
4	54 MPa
5	68 MPa
6	82 MPa
7	95 MPa
8	109 MPa

Figure 9.21 Stress distribution on the arm.

Table 9.1 Comparison of finite element analysis and experimental results

Position	Finite element analysis, MPa	Strain gauges, MPa	Photoelastic, MPa
Ball-joint housing	165	176	176
Inner radius body mount (with bushes)	20	25	Not available
Inner radius body mount (without bushes)	Not available	22	25

this comparison: the predictions are very good with only small errors between the computational and experimental results. From these results the validity of the finite element method for this case can be demonstrated, and now further work is ongoing to determine its validity in other more demanding circumstances.

9.4 DISPLACEMENT ANALYSIS OF A VEHICLE BODY SHELL

9.4.1 The Problem

In this final example a simple displacement analysis will be investigated for the monocoque structure of a typical saloon vehicle. Such an analysis is desirable as these structures are designed to have a given stiffness in both bending and torsion. High values of stiffness within the body shell enable accurate assembly to be carried out on the production line and are necessary for the vehicle to have good ride and handling qualities when being driven. Further, a high stiffness reduces the possibility of so-called NVH (noise, vibration and harshness) problems. These NVH problems include the amplification by the body shell of noise generated by the tyres moving over the road surface and by the air moving over the vehicle as well as the noise transmitted from the engine itself, together with extraneous noises caused by the resonant vibration of parts of the vehicle such as the exhaust or loose fittings. On the other hand, if the stiffness is low, not only will the ride, handling and NVH qualities be poor, but also other problems can arise such as vehicle doors not opening when the vehicle is parked with one wheel on a kerb or the front or rear window glass cracking or being forced out of the weather seals.

As with all engineering design problems the solution may well have to be a compromise. High cost or profit vehicles such as luxury saloons require excellent ride, handling and NVH properties and hence high stiffness in the structure. Conversely, lower cost or profit vehicles such as volume production hatchbacks and small saloons have lower standards applied to them in terms of structural stiffness as the functional requirements are much lower.

9.4.2 Building a Model of the Vehicle Body

Within the vehicle design process, external details such as the shape of the vehicle are the first things to be decided. This is done by the styling team for the vehicle.

Once the shape is decided then engineers take this shape, probably from a full-size clay model of the proposed vehicle, and produce a CAD model of the external panels. This is used to drive the design of the internal structure of the body shell. At this point a first-shot analysis to predict the stiffness of the body shell can be made.

A coarse finite element mesh is built from the CAD model, taking into account the known lines of the pressed steel panels and information based on experience, which gives a feel for the critical cross-sections where nodes must be placed for a successful analysis to be carried out. Figure 9.22 shows a coarse mesh in its displaced shape for a typical vehicle thin shell structure. This mesh consists of some 5000 linear triangular and quadrilateral elements, and the full body shell is modelled. The mesh has been created manually from the CAD data, but current practice is to use either a fully automatic mesh generation procedure or a combination of automatic and manual mesh generation methods. Note that the mesh models the vehicle roof and supporting pillars (known as the A, BC and D posts from front to rear), the main external steelwork at the rear (i.e. rear wings), the boot floor and internals, the floor pan and the structural steel around the engine (i.e. front bulkhead, suspension mountings and main beams).

Material properties are applied to the elements and in this case data for a typical steel has been used. Next the boundary conditions can be applied. This is done by restraining the geometric location of the body shell where the rear axle is attached to the body shell by the suspension mounts as shown by the double-headed arrows in Fig. 9.23. This simple schematic illustrates the application of the boundary conditions. Then equal and opposite vertical forces of magnitude F are applied at the front suspension mounting points, which are a distance L apart, as shown in Fig. 9.23. Hence a torque of magnitude FL is applied to the body shell. Note that these boundary conditions simulate the actual physical tests that are carried out on the body shell when the first prototypes are manufactured, and that the force F need only be a nominal force as the linear model will generate displacements which are proportional to the force.

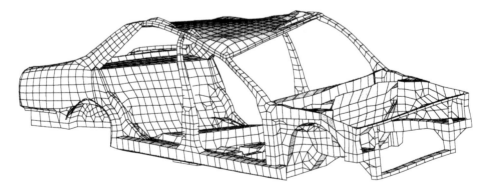

Figure 9.22 Displaced mesh for the vehicle body shell.

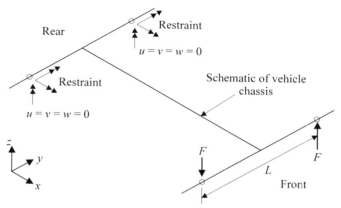

Figure 9.23 Loads applied to the vehicle body shell.

9.4.3 Analysing the Results for the Vehicle

The key result that is obtained from the analysis of this coarse model is the twist of the body shell per unit length of the body per unit loading. From experience this simple quantity can be used to decide if the actual body shell will have an acceptable level of stiffness. Figure 9.22 shows the displaced shape for this case, with the displacements much magnified of course. Note that the front offside of the roof is displaced downwards whereas the rear is displaced upward, and note that the opposite happens at the nearside.

This mesh is too coarse for the stress results to be of much use, but the analyst can use the stress data to identify key areas where high stresses may be expected. Either a more detailed model can be built of the whole structure or a model of some part of the structure (a submodel) can be built and analysed. This is another example of the use of substructuring as was discussed in Sec. 8.2.

A more detailed mesh is seen in Fig. 9.24, where a crash simulation has been carried out. This involves a nonlinear analysis of the structure, which will be dealt with in Sec. 10.4.

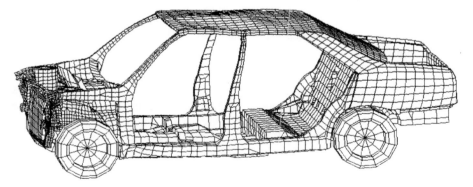

Figure 9.24 A simulation of a vehicle crash.

10

SOLVING MORE COMPLEX PROBLEMS

Throughout this book there has been a concentration on the development and application of the finite element method to linear static problems. In these   problems complex geometrical structures can be analysed to find the small   displacement of the structure from the applied loads when the material properties are such that nonlinear effects are negligible and effects over time are not important. While the vast majority of analyses are of this type, increasing use is being made of the finite element method in solving more complex problems. Typical examples are:

- the calculation of the variation in some property, say displacement, over time
- the determination of the mode shapes and frequencies of a vibrating structure
- the inclusion of optimization within a finite element program to allow, for example, either the shape of a structure or the thickness of shells to be modified to minimize the displacement of the structure at some point or to minimize the stress within the structure
- the treatment of material properties in the nonlinear range, such as during metal-forming processes when gross plasticity of the material occurs
- the linking of structural solvers to heat transfer solvers so that the effects of heating a structure in terms of both displacement and stress can be calculated.

This chapter is not intended to be a comprehensive guide to the fundamentals and applications of the finite element method to these problems. Rather, the intention is to give a brief overview of the ways in which such problems are handled, concentrating on the changes to the modelling process that occur which are different from those that have been outlined in previous chapters. Note that because the problems are handled differently to linear static problems, the modelling requires specifying additional parameters which the analyst will need to understand. Such modelling issues are covered in detail in the texts listed in Sec. 10.6.

10.1 TRANSIENT PROBLEMS

When considering the time-dependent behaviour of a structure, the motion is considered over a series of steps in time, with the solver calculating the structural motion from step to step. Here, the motion of the structure is described by considering the motion of the degrees of freedom in the finite element model. To model the necessary time-dependent terms describing the motion, i.e. the velocity and acceleration of the structure, a so-called *semi-discrete* form of the equations is produced from approximations to the time derivatives. For example, the first derivative of displacement with time can be approximated by using a truncated Taylor series expansion as

$$\left\{\frac{\partial \boldsymbol{\delta}}{\partial t}\right\} = \frac{\{\boldsymbol{\delta}\}^{n+1} - \{\boldsymbol{\delta}\}^n}{\Delta t} \tag{10.1}$$

where the superscripts denote time levels labelled $n + 1$ and n and Δt is the time interval. Use of shape functions and nodal values convert this, for an element, to

$$[\mathbf{N}]^{\mathrm{T}}[\mathbf{N}]\frac{\{\boldsymbol{\delta}^{\mathbf{e}}\}^{n+1} - \{\boldsymbol{\delta}^{\mathbf{e}}\}^n}{\Delta t} \tag{10.2}$$

When substituted in the governing energy equation for the problem, this leads to a new matrix $[\mathbf{M}^{\mathbf{e}}]$, which is known as the *element mass matrix*, where

$$[\mathbf{M}^{\mathbf{e}}] = \int_{V^e} \rho[\mathbf{N}]^{\mathrm{T}}[\mathbf{N}]\mathrm{d}V \tag{10.3}$$

where ρ is the density of the material of the element.

When this is integrated over an element, either the full integration can be performed to give the consistent mass matrix or approximations can be made to give a diagonal matrix, the lumped mass matrix. In (10.1) the value Δt acts to control the step-by-step procedure just as the relaxation parameter ω is used for the case of material nonlinearity. If Δt is very small the solution progresses slowly and convergence is more likely, whereas if Δt is large the time terms have very little effect and divergence is more likely.

As well as the time-dependent terms, those terms that are independent of time must also be discretized across various time levels. Often they are discretized just as they are for steady state problems, with the assumption that they are calculated at time level n. This leads to an explicit formulation of the equations which does not require the use of a simultaneous equation solver to produce a solution. Equally, though, the time-independent terms could be calculated at time level $n + 1$, leading to a fully implicit form of the equations. Also, some form of weighted average can be performed between these two extremes as follows:

$$\{\boldsymbol{\delta}\} = \theta\{\boldsymbol{\delta}\}^{n+1} + (1 - \theta)\{\boldsymbol{\delta}\}^n \tag{10.4}$$

Now setting θ to unity gives a fully implicit form and setting θ to zero gives the explicit form. For all other values an implicit form is produced and with θ set to 0.5 the Crank–Nicholson form is produced. If θ is 0.5 or greater then the schemes are unconditionally stable, and when θ is less than 0.5 they are conditionally stable. In the conditionally stable cases the time step Δt must be set small enough for stable solutions to be produced.

As far as the analyst is concerned, the following features need to be taken into account when a time-dependent simulation is produced:

- *The value of θ that is to be used* Typically 1.0 or 0.5 are used to ensure stability regardless of the time step.
- *The value of the time step Δt* This is set to provide stability if necessary when θ is less than 0.5, and must provide the analyst with a sufficiently detailed time response of the structure. This means that it must be small enough to give an accurate picture of the response.
- *The initial conditions* As the time terms have been discretized in this way a step-by-step procedure is involved that relies on data at time level $n + 1$ being produced from data at time level n. Hence to start the calculation at the first time step the values of the variables at time level zero must be known. These might be known from the specification of the problem or they may have to be assumed from experience.
- *The computing effort required* When any iterative procedure is involved, due to time variation or nonlinearity, the computer effort is very important as, effectively, the step-by-step procedure can easily be equivalent to many hundreds of individual static analyses.
- *The storage of data* It may be that the displacements of just a handful of nodes with time are required as the results of the simulation, but it is more likely that the displacement of all nodes at the different time steps will be required so that an animation of the structural response with time can be produced. For complex geometries such as a car body the number of degrees of freedom can be in the tens of thousands and storing this amount of data at several hundred time steps can require vast amounts of secondary disk storage.

10.2 VIBRATION PROBLEMS

Typical governing (or global) matrix forms of the time-dependent equation are:

$$[\mathbf{M}]\left\{\frac{\partial^2 \boldsymbol{\delta}}{\partial t^2}\right\} + [\mathbf{C}]\left\{\frac{\partial \boldsymbol{\delta}}{\partial t}\right\} + [\mathbf{K}]\{\boldsymbol{\delta}\} = \{\mathbf{F}(t)\} \tag{10.5}$$

where $[\mathbf{M}]$ is the mass matrix, $[\mathbf{C}]$ is the damping matrix and $\{\mathbf{F}(t)\}$ is the forcing function due to the load which may vary with time.

If a free vibration problem is considered, then there is no damping and no forcing of the structure, and so both $[\mathbf{C}]$ and $\{\mathbf{F}(t)\}$ are zero, giving:

$$[\mathbf{M}]\left\{\frac{\partial^2 \boldsymbol{\delta}}{\partial t^2}\right\} + [\mathbf{K}]\{\boldsymbol{\delta}\} = 0 \tag{10.6}$$

This situation is similar to the free vibration of a simple spring–mass system with one degree of freedom, which has a natural frequency given by

$$\omega_{\text{nat}} = \sqrt{\frac{k}{m}} \tag{10.7}$$

where k is the spring stiffness and m is the attached mass. If the general solution for the motion of the structure can be described in terms of nodal amplitudes $\{\bar{\delta}\}$ and a frequency ω, then

$$\{\delta\} = \{\bar{\delta}\}e^{i\omega t} \tag{10.8}$$

If this is substituted into (10.6), then

$$([\mathbf{A}] - \omega^2[\mathbf{I}])\{\bar{\delta}\} = 0 \tag{10.9}$$

where $[\mathbf{A}]$ is given by $[\mathbf{M}]^{-1}[\mathbf{K}]$ and is known as the *system matrix*. This is a classical eigenvalue problem. Solution of (10.9) for the eigenvalues gives n values of ω^2. Here n is the dimension of the symmetric matrices $[\mathbf{M}]$ and $[\mathbf{K}]$ and the values of ω are the natural frequencies of the system, each of which is associated with an eigenvector of the displacements of the structure. In fact, each eigenvector is a mode of vibration of the structure associated with the corresponding eigenvalue which is the frequency of vibration.

To deal with problems which have many thousands of degrees of freedom, one approach is to choose a limited number of degrees of freedom, known as master degrees of freedom. Through the use of mathematical manipulation the dynamics of the structure can be approximated by the response of these master degrees of freedom.

If damping is present, the damped natural frequency is obtained from

$$\omega_{\text{d}} = \omega_{\text{nat}} \sqrt{(1 - \xi^2)} \tag{10.10}$$

where ξ is the damping factor. It is often the physical situation that a structure is lightly damped with ξ less than 0.1. Under these conditions ω_{d} is within one-half of one per cent of ω and so the frequencies determined from equation (10.9) can be used to design against resonance in the damped structure. This is done by checking the size of the forcing frequencies to ensure that they are well away from the resonant frequencies. Further, the possible forms of any resonance can be determined from the mode shapes. Typical practical examples include the resonance of a cantilever beam structure such as a turbine blade or the resonance of a vehicle body shell.

10.3 OPTIMIZATION

Within the design process engineers have to work to produce a design that meets a given specification. Usually, there are many design solutions that meet the specification and so some sort of decision process must be undertaken to decide which solution is the *best solution* in some sense. Depending on the situation, determiners such as the cost of a product or the weight of a product may be used to evaluate the range of design solutions, with the so-called best solution being the one with the minimum value of cost or weight. Examples of this include the weight of an aircraft structure which is often minimized or the material cost of a plastic seat.

When the best solution is sought, the process is one of optimization. In this section this process will be considered in broad terms, defining some vocabulary, looking at search techniques and then considering how the process fits in with the finite element method.

10.3.1 Vocabulary

The first term that needs a definition is *optimization*. This is defined as the mathematical process of obtaining the set of conditions required to produce the maximum or minimum value of some function. Within the problem, *design variables* are found which are the quantities that can be used to define a product but are not predefined for the problem being considered. These variables may be the thickness of some material, the weight of the product or its cost and so on. One of these variables is minimized in the optimization process. Each problem may also be subject to *constraints* which, typically, are derived from known limits for, say, displacement or stress values in a material.

The entity which is optimized is known as the *objective* or *cost function*, which is a functional form of one of the design variables in terms of the other design variables. For example, if the objective function is D and the other design variables are x_1 to x_n, then

$$D = f(x_1, x_2, \ldots, x_n) \qquad (10.11)$$

10.3.2 Seeking a Minimum

If it is considered that seeking the maximum value of a function is the same as seeking the minimum value of the negative of the function, then all optimization can be seen as a minimization process. Differentiation of (10.11) to find the design variables that give a zero value for the derivative may be one way to proceed. However, for complex structural problems, where (10.11) can only be found numerically as there are a large number of design variables, this task is virtually impossible. Numerical methods are, therefore, used to seek the minimum.

The simplest way is the *direct search method* where the value of D is found for a number of values of each of the design variables x_1 to x_n. However, a large number of calculations may be required to do this and so, for multi-variable problems, *pattern search* methods can be used. These evaluate D at a given set of design variables, known as the *base point*, and then consider just one design variable, evaluating D either side of the original position. Where D is a minimum, for the three points being considered, a new local search in a second design variable is made and so on. When all the variables have been considered, the local minimum is found and the process can be repeated from some new base point. This new point is found by moving away from the original base point in the direction of the local minimum by, say, twice the distance between original base point and local minimum. Other methods can also be used and these involve finding the steepest gradient of D at the base point and moving in that direction before re-evaluating the gradients.

10.3.3 The Finite Element Context

When using finite elements for structural design, it is possible to use a manual form of optimization in conjunction with a series of finite element analyses. For example, say an analyst wishes to know the minimum thickness of material that has to be used within a shell structure to achieve a given displacement at some point for a given loading subject to the material not being stressed beyond some value. A finite element shell model can be built and the thickness set to some arbitrary value. Then a linear static analysis can be run and the displacements and stresses can be found. Assuming that the stress constraints are not violated, if the maximum displacement is too large then the thickness must be increased, and if it is too small then the thickness can be reduced. The thickness might be halved or doubled as appropriate and the simulation run again. Repeating this process should lead to the optimum thickness.

However, programs are now available that automate this process. First of all, the relevant design variables must be defined and their initial values set. Then a finite element analysis is performed to calculate the structural behaviour and generate values for the constraints. A so-called *sensitivity analysis* is then performed that calculates the rate of change of the structural behaviour and the rate of change of the constraints with the design variables. From the constraint values and their derivatives a localized model of the constraints as a function of the design variables can be built and a search carried out to improve the design variables such that the cost function is minimized. These design variables are then updated within the finite element model and a second analysis is performed. Repeating the full process results in an optimum design.

10.4 NONLINEAR PROBLEMS

10.4.1 Nonlinearities in the Material Properties

Here the material properties of the problem lead to what is, at first sight, a small change in the finite element equation. Equation (1.2) becomes:

$$[\mathbf{K}(\boldsymbol{\delta})]\{\boldsymbol{\delta}\} = \{\mathbf{F}\} \tag{10.12}$$

which signifies that the global stiffness matrix $[\mathbf{K}]$ is now a function of the nodal displacements $\{\boldsymbol{\delta}\}$. Here the dependency of the terms in $[\mathbf{k}]$, as given by (3.9), is because the elastic stiffnesses C_{ij} in $[\mathbf{D}]$ (2.15) are functions of strain. This change has enormous repercussions when trying to produce a solution to (10.12), as the actual nodal displacements are not known until some complex solution procedure is complete.

To produce a solution, an iterative procedure must be used. To start this procedure, the analyst must provide some guessed values for the displacements to the solver program. These initial values are then used by the solver to calculate an approximate set of coefficients for the matrix $[\mathbf{K}]$. Having done this any of the the usual linear equation solvers (see Sec. 3.9.3) can be used to solve this initial version of (10.12). This yields a new set of values for the displacements and step one of the process is complete. These updated values can then be used within step two of the process as new estimates for the nodal displacements and the coefficients of matrix $[\mathbf{K}]$ are calculated, taking account of the changes in the material properties as calculated at the current element stresses and strains (see Sec. 2.2.6). Further solution of (10.12) gives another set of values for the displacements at the end of step two. Further steps in the process are carried out such that the process is repeated many times until the calculated values of the displacement do not change from one step to the next.

Owing to the iterative nature of the process, the displacements may not converge to a given set of values from step to step. Sometimes the displacements can oscillate wildly from step to step, with the magnitude ever increasing in value. When this divergence happens the process is unstable and usually the computer detects that the numbers being generated are outside its range and it stops the process automatically. To assist the process to converge, relaxation techniques can be used where the displacements that are used in the next step are calculated using

$$\{\boldsymbol{\delta}_{\text{next}}\} = \omega\{\boldsymbol{\delta}_{\text{new}}\} + (1 - \omega)\{\boldsymbol{\delta}_{\text{old}}\} \qquad (10.13)$$

where the relaxation parameter ω can range from zero to unity. Hence, when ω is 1.0 the iteration process takes place as originally described, with the displacements being taken as the newly calculated values in the next step, and when ω is 0.0 the iteration process is stopped as the displacements are never updated. Taking some intermediate value of ω increases the likelihood of convergence where this is necessary, by effectively slowing down the iterative process.

For the analyst the implications of attempting to calculate a nonlinear problem can be immense. Common problem areas include the following:

- *Provision of a guessed set of values for the displacement*   If sensible choices are made then the convergence of the process can be improved, but if poor choices are made then divergence is likely. Experience of running similar problems can be a great advantage here.
- *Monitoring the iterative process for divergence*   It is useful to check the value of displacement, say at one point, at the end of every step. If the displacements are not oscillating and the change in value from step to step after, say, 10 steps is reducing then the process is probably converging.

- *Setting the number of steps to be run during the iteration process* Here it is sensible to run just a few steps to start with. This enables monitoring to be carried out and convergence to be checked before calculating many more steps. Often several hundred steps may be needed. Usually the solver program allows a restart of the solution from data that has been saved to disk at the end of the last step. This saves rerunning previous steps in the process as the procedure progresses.
- *The choice of relaxation parameter ω* Again, experience of running similar problems may well lead to a sensible choice of ω.
- *The increase in computer time over a single linear static calculation* It is now clear that many equivalent linear static solutions are calculated during the iterative procedure. Hence the computer effort is magnified by the number of steps run. This may mean that a smaller model in terms of the numbers of nodes and elements may have to be built than would be ideal, in an attempt to solve the problem with a realistic amount of computer effort. It is this increase in computing requirement that has led to an increase in the use of iterative linear equation solvers. While these may be more inaccurate than direct solvers they are much faster. Also, during the initial stages of the iterative process, an inaccurate solution for the displacements is acceptable.

In many ways structural problems that are nonlinear are similar to nonlinear problems in the field of computational fluid dynamics. Analysts in computational fluid dynamics who have to solve problems similar to structural nonlinear problems might find the detailed modelling advice in Shaw (1992) useful.

10.4.2 Geometrical Nonlinearities

With the material nonlinearity problems described above the displacements and strains in a structure may still be regarded as small. However, a second type of nonlinearity, geometrical nonlinearity, is also possible where the displacements are large, and this type nonlinearity is also defined by (10.2). Geometric nonlinear behaviour is introduced into the finite element method by including the third term of a Taylor series expansion in the definition of strains (see Sec. 2.2.4) and constructing the [**k**] using the same procedure as in Sec. 3.1.3. A simple example of this behaviour is a buckled strut with imperfections. As a general rule, most buckling problems should be approached as nonlinear problems in which pre-buckling deformations are taken into account. In these problems it is also possible for the material nonlinearity to be present as well.

Buckling can also be solved using an eigenvalue analysis similar to that for free vibration. This type of analysis is often available in packages.

Analysis by the finite element method of problems such as metal forming in a press and impact response of, say, road vehicles requires a combination of material and geometrical nonlinearities. These problems are therefore the most difficult to model and solve with accuracy.

10.5 CALCULATING THERMAL EFFECTS WITHIN STRUCTURES

It is often required that the effects of heating or cooling a structure on its stress and strain distribution need to be known. The thermal problem can be, effectively, decoupled from the stress–strain problem through the use of an initial stress and strain field as shown in Chapter 3.

To solve the thermal problem, equations of the form

$$\frac{\partial^2 T}{\partial x^2} + \frac{\partial^2 T}{\partial y^2} + \frac{\partial^2 T}{\partial z^2} + Q = 0 \tag{10.14}$$

are solved. This is Poisson's equation where $T(x, y, z)$ is the local temperature and $Q(x, y, z)$ is a local source of heat. A weighted residual form of this equation is easily formed by multiplying the equation by some test function and then by integrating by parts over the volume of an element. By using shape functions and nodal temperatures, a finite element form of this is easily produced:

$$-\int_V \left[\left[\frac{\partial \mathbf{N}}{\partial x}\right]^T \left[\frac{\partial \mathbf{N}}{\partial x}\right] + \left[\frac{\partial \mathbf{N}}{\partial y}\right]^T \left[\frac{\partial \mathbf{N}}{\partial y}\right] + \left[\frac{\partial \mathbf{N}}{\partial z}\right]^T \left[\frac{\partial \mathbf{N}}{\partial z}\right] + [\mathbf{N}]^T [\mathbf{N}] \right] dV + \int_S [\mathbf{N}]^T \frac{\partial T}{\partial n} dS = 0 \tag{10.15}$$

Here V denotes the volume of the problem and S denotes the surface boundary. At the boundary either the temperature T or the normal derivative of temperature can be the specified boundary conditions. In fact, one nodal value of T must be given to ensure that (10.15) is not singular. Given a mesh of elements and the boundary conditions, the value of temperature can be found from (10.15) throughout the structure. These can then be converted to initial strains (see Sec. 3.1.1) within the structure and a new calculation can be carried out to find the combined stresses and strains when external loads are applied as well. If the temperature distribution is not constant, then the elastic constants will vary through the body and affect the structural behaviour as defined in (10.12).

10.6 REFERENCES AND FURTHER READING

Detailed technical information on the fundamental principles are given in texts such as those by Cook *et al.* (1989), Crisfield (1991), Hinton (1992), Hitchings (1992), Morris (1982) and Zienkiewicz and Taylor (1991). Modelling issues are also discussed in NAFEMS (1986) and Cook (1995). Both fundamental principles and modelling issues are to be found in journals such as: *Computers and Structures*, Pergamon Press; *Finite Element in Analysis and Design*, Elsevier Science BV; *International Journal for Numerical Methods in Engineering*, John Wiley.

REFERENCES AND FURTHER READING

Ahmad, S., Irons, B. M. and Zienkiewicz, O. C. (1970) Analysis of thick and thin shell structure by curved finite elements, *Int. J. Numer. Methods Engng.*, vol. 2, 419–451.

Airy, G. B. (1863) On the strains in the interior of beams, *Phil. Trans. Roy. Soc.*, vol. 153, 49–80.

Argyris, J. H. and Kelsey, S. (1960) *Energy Theorems and Structural Analysis*, Butterworth, London.

Ashby, M. F. (1992) *Materials Selection in Mechanical Design*, Pergamon, London.

Bézier, P. (1986) *The Mathematical Basis of the UNISURF CAD System*, Butterworth, London.

Burnett, D. S. (1987) *Finite Element Analysis From Concepts to Applications*, Addison-Wesley, Reading, Mass.

Calladine, C. R. (1983) *Theory of Shell Structures*, Cambridge University Press, Cambridge.

Calladine, C. R. and Christopher, R. (1985) *Plasticity for Engineers*, Ellis Horwood, Chichester, England.

Cavendish, J. C., Field, D. A. and Frey, W. H. (1985) An approach to automatic three-dimensional finite element mesh generation, *Int. J. Numer. Methods Engng.*, vol. 21, 329–47.

Charles, J. A. and Crane, F. A. A. (1989) *Selection and Use of Engineering Materials*, 2nd edn, Butterworth, London.

Cheng, J. H., Finnigan, P. M., Hathaway, A. F., Kela A. and Schroeder, W. J. (1988) Quadtree/Octree meshing with adaptive analysis, in Senegupta, S., Thompson, J. F., Eiseman, P. R. and Hauser, J. (eds.), *Numerical Grid Generation in Computational Fluid Dynamics*, Pineridge Press, Swansea, pp. 633–42.

Clapeyron, B. P. E. and Lamé, G. (1833) *Mémoire sur L'équilibre Intérieur des Corps Solides Homogènes*, Mem. Savans Etrang, Paris, IV, 463–562.

Clough, R. W. (1960) The finite element in plane stress analysis, *Proc. 2nd ASCE Conf. on Electronic Computation*, Pittsburgh, Pa., September 1960.

Coates, R. C., Coutie, M. G. and Kong, K. F. (1988) *Structural Analysis*, 3rd edn, Van Nostrand Reinhold, Wokingham.

Cook, R. D. (1995) *Finite Element Modelling for Stress Analysis*, Wiley, New York.

Cook, R. D., Malkus, D. S. and Plesha, M. E. (1989) *Concepts and Applications of Finite Element Analysis*, 3rd edn, Wiley, New York.

Courant, R. (1943) Variational methods for the solution of problems of equilibrium and vibration, *Bull. Am. Math. Soc.*, vol. 49, 1–23.

Crisfield, M. A. (1991) *Non-linear Finite Element Analysis of Solids and Structures, Vol. 1: Essentials,* Wiley, New York.

Cross, H. (1930) The analysis of continuous frames by distributing fixed-end moments. *Proc. Amer. Soc. Civ. Engrs.,* vol. 56, 919–28.

Cuthill, E. and McKee, J. (1969) Reducing the bandwidth of sparse symmetric matrices, *Proc. 24th ACM National Conference,* New York, pp. 157–72.

Desai, C. S. and Abel, J. F. (1972) *Introduction to the Finite Element Method,* Van Nostrand Reinhold, New York.

Deutschman, A. D., Mitchels, W. J. and Wilson, C. (1975) *Machine Design,* Macmillan, New York.

Dieter, G. E. (1986) *Mechanical Metallurgy,* 3rd edn, McGraw-Hill, New York.

Donnell, L. H. (1933) Stability of thin-walled tubes under torsion. *NACA Report 497,* National Advisory Committee for Aeronautics, Washington D.C.

Encarnacao, J. and Schlechtendahl, E. G. (1983) *Computer Aided Design—Fundamentals and System Architectures,* Springer-Verlag, Berlin.

Fenner, R. T. (1986) *Engineering Elasticity: Applications of Numerical and Analytical Techniques,* Ellis Horwood, Chichester, England.

Flügge, W. (1973) *Stresses and Strains in Shells,* 2nd edn., Springer-Verlag, Berlin.

Galerkin, B. G. (1915) Series solution of some problems of elastic equilibrium of rods and plates (in Russian), *Vestn. Inzh. Tech.,* vol. 19, 897–908.

George, P. L. (1991) *Automatic Mesh Generation—Application to Finite Element Methods,* Wiley, Chichester (Masson, Paris).

Gere, J. M. and Timoshenko, S. P. (1991) *Mechanics of Materials,* 3rd edn, Chapman & Hall, London.

Gibbs, N. E., Poole, W. G. and Stockmeyer, P. K. (1976) An algorithm for reducing the bandwidth profile of a sparse matrix, *SIAM J. Numer. Anal.,* vol. 13(2), 236–50.

Gordon, J. E. (1976) *The New Science of Strong Materials,* 2nd edn, Penguin, Harmondsworth, England.

Halpin, J. C. (1992) *Primer on Composite Materials Analysis,* 2nd edn, Technomics, Lancaster, Pa.

Hinton, E. (ed.) (1992) *NAFEMS Introduction to Non-linear Finite Element Analysis,* NAFEMS, Glasgow.

Hinton, E. and Owen, D. R. J. (1979) *An Introduction to Finite Element Computations,* Pineridge Press, Swansea.

Hirsch, C. (1988) *Numerical Computation of Internal and External Flows, Volume 1: Fundamentals of Numerical Discretisation,* Wiley, New York.

Hitchings, D. (ed.) (1992) *A Finite Element Dynamic Primer,* NAFEMS, Glasgow.

Hoffman, J. D. (1992) *Numerical Methods for Engineers and Scientists,* McGraw-Hill, New York.

Holmes, D. G. and Lanson, S. H. (1986) Adaptive triangular meshes for compressible flow solutions, in Hauser, J. and Taylor, C. (eds.), *Numerical Grid Generation in Computational Fluid Dynamics,* Pineridge Press, Swansea, pp. 413–24.

Hordeski, M. F. (1986) *CAD/CAM Techniques,* Reston Pub. Co., Reston, Va.

Hrenikoff, A. (1941) Solution of problems in elasticity by the framework method, *J. Appl. Mech.,* vol. A8, 169–75.

Hughes, T. J. R. and Hinton, E. (1986) *Finite Element Methods for Plate and Shell Structures. Vol. 1 Element Technology,* Pineridge Press, Swansea.

Irons, B. M. (1966) Engineering applications of numerical integration in stiffness methods, *AIAA J.,* vol. 4(11), 2035–7.

John, V. (1992) *Testing of Materials,* Macmillan, Basingstoke.

Kaye, G. W. C. and Laby, T. H. (1983) *Tables of Physical and Chemical Constants,* 14th edn, Longman, London.

Knight, C. E. (1993) *The Finite Element Method in Mechanical Design,* PSW-Kent, Boston, Mass.

Lamit, L. G. (1994) *Technical Drawing and Design,* West, St Paul, USA.

Levy, S. (1953) Structural analysis and influence coefficients for delta wings, *J. Aero. Sci.,* vol. 20(7), 449–54.

Lewis, P. E. and Ward, J. P. (1991) *The Finite Element Method—Principles and Applications,* Addison-Wesley, Wokingham.

Love, A. E. H. (1944) *A Treatise on the Mathematical Theory of Elasticity*, 4th edn, Dover.

Martin, H. C. and Carey, G. F. (1973) *Introduction to Finite Element Analysis, Theory and Applications*, McGraw-Hill, New York.

Maxwell, J. C. (1872) On reciprocal figures, frames, and diagram of forces, *Edinburgh Roy. Soc. Trans.*, vol. XXVI, Edinburgh, 1–40.

McHenry, D. (1943) A lattice analogy for the solution of plane stress problems, *J. Inst. Civ. Engrs.*, vol. 21, 59–82.

McMahon, C. and Browne, J. (1993) *CADCAM From Principles to Practice*, Addison-Wesley, Wokingham.

Mindlin, R. D. (1951) Influence of rotary inertia and shear on flexural motions of isotropic, elastic plates, *J. Appl. Math.*, vol. 18, 31–8.

Morris, A. J. (ed.) (1982) *Foundations of Structural Optimization: A Unified Approach*, Wiley, Chichester, England.

NAFEMS (1986) *A Finite Element Primer*, NAFEMS, Glasgow.

NAFEMS (1992) *Benchmark*, NAFEMS, Glasgow (from 1988 to present).

Norrie, D. and de Vries, G. (1976) *Finite Element Bibliography*, IFI/Plenum, New York.

Onwubiko, C. (1989) *Foundations of Computer-Aided Design*, West, St Paul, USA.

Packer, J. A., Wardenier, J., Kurobane, Y., Dutta, D. and Yeomans, N. (1992) *Design Guide for Hollow Rectangular Section (RHS) Joints Under Predominantly Static Loading*, CIDECT ed. Verlag TUV Rheinland, Germany.

Pahl, G. and Beitz, W. (1988) *Engineering Design, A Systematic Approach* (Trans. A. Pomerans and K. Wallace), The Design Council, London.

Pinfold, M. and Calvert, G. (1994) Experimental analysis of a composite automotive suspension arm, *Composites*, vol. 25(1), 59–63.

Rayleigh, J. W. S., Lord (1870) On the theory of resonance, *Trans. Roy. Soc. (London)*, vol. A161, 77–118.

Reddy, J. N. (1984) *An Introduction to the Finite Element Method*, McGraw-Hill, New York.

Richardson, L. F. (1910) The approximate arithmetical solution by finite differences of physical problems, *Trans. Roy. Soc. (London)*, vol. A210, 305–57.

Ritz, W. (1909) *Uber eine neue Methode zur Lösung gewissen Variations—Probleme der mathematischen Physik*, *J. Reine Angew. Math.*, vol. 135, 1–61.

Robinson, J. (1985) *Early FEM Pioneers*, Robinson & Associates, Dorset, England.

Rooney, J. and Steadman, P. (1987) *Principles of Computer-aided Design*, Pitman and The Open University, London.

Shaw, C. T. (1992) *Using Computational Fluid Mechanics*, Prentice Hall, London.

Shigley, J. E. and Mitchell, L. D. (1983) *Mechanical Engineering Design*, McGraw-Hill, New York.

Smith, G. D. (1985) *Numerical Solution of Partial Differential Equations: Finite Difference Methods*, 3rd edn, Oxford University Press, Oxford.

Southwell, R. V. (1940) *Relaxation Methods in Engineering Science: Treatise on Approximate Computation*, Oxford Clarendon Press, Oxford.

Stasa, F. L. (1986) *Applied Finite Element Analysis for Engineers*, CBS Publishing, New York.

Szmelter, J. (1958) The energy method of networks of arbitrary shape in problems of the theory of elasticity, *Proc. IUTAM Symposium on Non-homogeneity in Elasticity and Plasticity* (ed. Olszak, W.), Warsaw 2–9 September 1958, Pergamon Press, London.

Taig, I. C. (1992) *Finite Element Analysis of Composite Materials*, NAFEMS, Report R0003, Glasgow.

Taylor, C. and Hughes, T. G. (1981) *Finite Element Programming of the Navier–Stokes Equations*, Pineridge Press, Swansea.

Thompson, J. F., Warsi, Z. U. A and Mastin, C. W. (1982) Boundary-fitted coordinate systems for numerical solution of partial differential equations—A review, *J. Comput. Phys.*, vol. 47(1), 1–108.

Timoshenko, S. P. (1953) *History of Strength of Materials*, McGraw-Hill, New York.

Timoshenko, S. P. and Goodier, N. J. (1988) *Theory of Elasticity*, 2nd edn, McGraw-Hill, New York.

Tizzard, A. (1994) *An Introduction to Computer-aided Engineering*, McGraw-Hill, London.

Turner, M. J., Clough, R. W., Martin, H. C. and Topp, L. J. (1956) Stiffness and deflection analysis of complex structures, *J. Aero. Sci.*, vol. 23(9), 805–33.

Vlasov, V. Z. (1964) *General theory of shells and its applications to engineering.* (Translated from Russian in 1949.) Nasa Technical Translation TTF-99. US National Aeronautical and Space Administration, Washington D.C.

Washizu, K. (1982) *Variational Methods in Elasticity and Plasticity*, Pergamon Press, Oxford.

Watson, D. F. (1981) Computing the *n*-dimensional Delaunay tesselation with application to Voronoi polytypes, *Comput. J.*, vol. 24(2), 167–72.

Yerry, M. A. and Shephard, M. S. (1984) Automatic three-dimensional mesh generation by the modified-octree method, *Int. J. Numer. Methods Engng.*, vol. 20(11), 1965–90.

Young, W. C. (1989) *Roark's Formulas for Stress and Strain*, 6th edn, McGraw-Hill, New York.

Zienkiewicz, O. C. and Cheung, Y. C. (1967) *The Finite Element Method in Structural and Continuum Mechanics*, McGraw-Hill, New York.

Zienkiewicz, O. C. and Taylor, R. L. (1989) *The Finite Element Method, Volume 1: Basic Formulation and Linear Problems*, 4th edn, McGraw-Hill, New York.

Zienkiewicz, O. C. and Taylor, R. L. (1991) *The Finite Element Method, Volume 2: Solid and Fluid Mechanics Dynamics and Non-linearity*, 4th edn, McGraw-Hill, New York.

INDEX